FRED D. RAUCH
PAUL R. WEISSICH

PLANTS FOR THE TROPICAL XERISCAPE

A Gardener's Guide

UNIVERSITY OF HAWAI'I PRESS
HONOLULU

The production of this volume has been supported by generous donations from the following organizations and individuals:

BOARD OF WATER SUPPLY, CITY AND COUNTY OF HONOLULU

FRIENDS OF THE HALAWA XERISCAPE GARDEN

KAULUNANI URBAN AND COMMUNITY FORESTRY PROGRAM, DIVISION OF FORESTRY AND WILDLIFE, STATE OF HAWAIʻI

FRIENDS OF THE HONOLULU BOTANICAL GARDEN

AMERICAN SOCIETY OF LANDSCAPE ARCHITECTS

LANDSCAPE INDUSTRY COUNCIL OF HAWAIʻI

DONALD HODEL

Printed in China

20 19 18 17 16 15 6 5 4 3 2 1

Rauch, Fred D. (Fred Donald), author.

Plants for the tropical xeriscape : a gardener's guide / Fred D. Rauch and Paul R. Weissich.

pages cm

Includes bibliographical references and index.

ISBN 978-0-8248-4005-1 (cloth : alk. paper)

1. Xeriscaping. 2. Drought-tolerant plants. I. Weissich, Paul R., author. II. Title.

SB439.8.R37 2015 635.9—dc23 2014014863

University of Hawaiʻi Press books are printed on acid-free paper and meet the guidelines for permanence and durability of the Council on Library Resources.

DESIGNED BY JULIE MATSUO-CHUN

PRINTED BY REGENT PUBLISHING SERVICES, LTD.

CONTENTS

INTRODUCTION

In 1981 the Denver, Colorado, Water Department coined the word "xeriscape" to describe new gardening techniques to conserve water. The word is derived from a Greek word indicating dry conditions. The principles of xeriscaping quickly spread throughout the arid southwest and were especially embraced in the dry areas of southern California, resulting in significant water savings.

O'ahu has an ample supply of excellent water. But as the population grows, so does water usage. The supply does not. Water is a vital and finite resource. Population growth and tourist development on all the islands are greatest in the warm, dry, leeward sections. This is true of many tropical areas, especially high islands that "catch" rainfall on their windward slopes, resulting in a rain shadow (reduced rainfall) on their leeward sections. Eventually, probably quite soon, some form of water conservation will be required.

Half of O'ahu's water usage is residential. Of that, half is used to sustain the landscape. Factor in the landscape water usage of schools, parks, golf courses, government facilities, and the many hotel and resort properties on O'ahu, and the problem becomes far more complex. As much as an 80 percent reduction in landscape water usage can be realized through intelligent water management: through xeriscaping. This is true for all water users public and private. We are looking at a very substantial water saving potential.

It is important to note that almost half of the world's tropics are arid or partially arid, belying the traditional Hollywood depiction of the tropics as moist, lush, and green. Large areas of tropical Africa, Australia, India, Southeast Asia, and the New World are deserts, near deserts, or seasonal deserts. Even the garden island of Sri Lanka has its dry area. The old adage "Every cloud has a silver lining," however, is particularly true for the xeriscaper, for these dry tropics are the source of a wide variety of beautiful, landscape-worthy species that are also drought tolerant. Their use is an important element in the xeriscape.

Landscape architects, garden designers, and the home weekend gardener are challenged with creating a new tropical image, one with color, shade, and a new lushness. These attractive aspects are completely within reach of the thoughtful xeriscaper.

The principles of xeriscaping are simply stated. They call for setting up "zones"

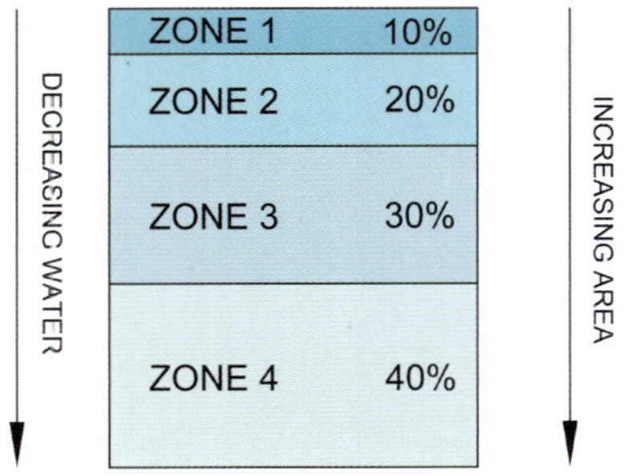

WATER CONSERVATION CHART

of water application: small areas for high water users, larger areas for more drought-resistant species, and even larger areas for species that are highly drought tolerant. The chart clarifies these relationships.

We have elected to use four zones, but the xeriscaper may use more or fewer. Zones may be of any shape.

Each zone is provided with its own water distribution system, which, following a period of establishment, is programmed to deliver just the amount of water required for the plants selected for that zone. Some trial and error is required to determine just how much water is required for zoned plants in your location. There is a highly technical process called *evapotranspiration* that provides this information, but it is not reasonably available for the general landscaper. See Appendix F: Evapotranspiration.

There are refinements that the gardener can make once water usage is determined. There are automated systems and ground moisture sensors that turn off the automation in the case of rain or already wet soil. Your local sprinkler vendor can be of great help if you purchase your equipment from him.

There are many good "how-to" books available that instruct the home gardener and professional in the design and installation of a successful xeriscape irrigation system (see the suggested readings section for examples). It is the intent of this publication, however, to lead the gardener to those landscape plant species zoned "less thirsty" that form the major part of the tropical xeriscape.

Plant names are grouped under several categories in accordance with their use in the xeriscape—such as ground covers, small shrubs, and large shrubs—and within each landscape use group, they are listed alphabetically by genus. For the most part, scientific names are those found in *A Tropical Garden Flora* by George W. Staples and Derral R. Herbst. Other taxonomic and descriptive sources are listed in the suggested readings. Every effort has been made to obtain the most current plant name information. Errors are the sole responsibility of the authors. Common names for plants vary widely throughout the world. Those most commonly used in Hawai'i, plus Hawaiian plant names where available, have been selected. Where no common name was found, we have taken the liberty of suggesting one.

To facilitate plant selection, one may briefly summarize the principles as follows: develop a plan that will locate which garden spaces are needed for minimal turf areas (which are heavy water users), planted spaces, paved surfaces, and the subsequent design of an efficient irrigation system that is specifically programmed to deliver lessening amounts of water to each planting space.

The plan can also guide grading: construct broad, shallow basins around major plantings to prevent runoff and insure the greatest water penetration; slightly tilt paved surfaces into water retention areas. Collect rainwater in a large container that can be used to water potted plants.

The plan will permit easy "zoning" of the garden: a small space for plants requiring high water use, a larger space for species needing low water applications, and one or more major spaces for those landscape species requiring minimal watering. The landscape may be zoned into even finer divisions. Zones may be of any shape, but some form of zoning is critical. Currently, gardeners usually apply water throughout the garden at a rate required by the highest water users. This results in overwatering and waste. Xeriscaping eliminates this waste.

Those not familiar with the xeriscape concept err in having visions of a ground cover of gravel or cinders and a planting of cactus. Nothing is further from the truth. A good xeriscape plan can accommodate ferns and other water lovers—even a water feature: a fountain or pond. But in a relatively small space, xeriscape!

It is important to note that even the least "thirsty" species will require a period of establishment, during which the root system spreads and the plant becomes adjusted to its new environment—assuming, of course, that the specimen was nursery grown or transplanted from another garden. The period of establishment will vary for each site, but think in terms of two years. Gradual reduction of applied water, usually during the short-day, rainy season, will attain the designed application without negative impact on the plantings. The actual final watering schedule is determined by each gardener from careful observation, adjustment, and trial and error based on the location and other factors noted below. There is no magic "rule of thumb." And remember: if new plants are introduced into a mature, balanced xeriscape, they must be hand-watered until they are also established and have adjusted to their new zoned locations.

The following listings are organized according to landscape use—ground covers, small shrubs, large shrubs, etc.—and are in alphabetical order. To clarify the frequent use of the term "moderately drought tolerant," we have assumed four zones of drought tolerance:

- ZONE 1 contains those species of high water use such as heliconias, hybrid ti, gingers, a lily pond, grassed areas, and ferns and is given the smallest area in the garden. We do not list any Zone 1 species, but this zone is an integral part of the xeriscape plan. Refer to the authors' earlier publications, *Plants for Tropical Landcapes* (2000) and *Small Trees for Tropical Landscapes* (2009).
- ZONE 2 lists plants of low drought tolerance.
- ZONE 3 lists species of significant drought tolerance and is given greater area than Zone 2.
- ZONE 4 lists species of high drought tolerance and is afforded the largest area of the garden.

Each species description is followed by its zone placement (e.g., Zone 2, 3, or 4). Appendixes A through E list plants in each zone under their landscape use. We hope to err on the side of conservatism in assigning zones.

It is necessary to keep in mind that these levels of drought tolerance are relative and not exact quantifiers. They also do not reflect mere survival but rather a tolerance level that permits the plant to retain acceptable visual quality for the garden. There are many species that shrivel, wilt, and change color with low water availability but return to normal with application of water. This is not visually acceptable.

Note that average annual rainfall and distribution of rainfall are important factors. Plants rooted in a deep soil will survive more dryness than those in shallow soils. The presence or absence of groundwater or a high water table is a critical factor. High humidity versus low humidity is a factor. Wind is also an important factor. A windy site will become dry far more rapidly than a protected site. The same is true for sunny versus shaded garden spaces.

It must also be noted that plants recommended in publications for xeriscaping in many tropical and subtropical areas may not be appropriate for Hawai'i. The bottom line is that each gardener must keep an open mind, plant with intelligence, and be willing to make changes as demonstrated by the health or demise of his specific plantings.

The most important mandate is that we give xeriscaping a chance. We must conserve water.

This publication is in compliance with the Hawaii-Pacific Weed Risk Management Project (see Appendix G) and the constraints listed by the Landscape Industry Council of Hawaii (LICH).

A word about our treatment of succulent plants is in order. Succulents have evolved in arid areas and have developed marvelous water-conserving mechanisms; special characters from tough, heat-reflecting surfaces to pubescence and hairs that protect the plant from wind and hot sun; oils that do not easily give up their moisture; spines, thorns, and toxic sap to inhibit grazing and browsing; and water-holding tissues that sustain the plant between periods of scarce rain. Many of these are among the most beautiful plants available to the xeriscaper. We have selected only a few from the myriad of available succulents—those which, in our view, are most noteworthy. The reader is directed to the suggested readings section, which lists publications that augment our limited selections.

On the whole, most gardeners are not cactus enthusiasts. Because there are numerous excellent illustrated publications dealing with the cactus family (Cactaceae) containing many well-known and highly ornamental, drought-tolerant tropical species, we have elected to omit these from our listing. One exception is the genus *Pereskia,* which presents unusual qualities. See the suggested readings section for publications dealing with the Cactaceae. Remember that most temperate zone cactus species will not grow successfully in a tropical climate. Select those from similar climates.

Do not hesitate to visit the well-presented shows put on by your local cactus and succulent societies. Your best information and advice is available from those generous enthusiasts.

A word to the wise gardener: plants produce a wide range of chemical substances as well as physical properties that must be considered in selecting plants for the xeriscape. Some have thorns, spines, irritating hairs, or even poisons that may create problems if not recognized and dealt with. Of major concern to parents are plants containing toxins that might cause serious medical problems. Reactions vary from person to person and from child to child depending on body weight and amount of exposure.

Teach your children not to smell, pick, or—especially—chew or swallow leaves, flowers, or fruit. Children are carefully taught to obey traffic lights, use the pedestrian crosswalk, and to follow strict safety rules while riding a bicycle. These activities are far more potentially dangerous than plant toxins. Those species in the text that may pose problems are marked as follows: **T** after the common name indicates the presence of thorns or spines; **S** indicates species known to have caused skin, eye, mouth, or throat irritation; **P** indicates plants with toxic sap, leaves, flowers, or fruit; and + indicates plants deserving special care in handling.

Additional information on xeriscaping and less thirsty species may be found in the suggested readings section of this book.

Nitrogen fixing is important. There are a number of plants capable of absorbing atmospheric nitrogen and converting it into an available form. This is a great advantage in a garden with poor soil. In this book, known nitrogen-fixing species are noted by the addition of **N** following the common name and **NP** for a species of the same genus that has a high probability of also being a nitrogen fixer.

ACKNOWLEDGMENTS

The authors are indebted to the following for providing special professional information or services: Dr. George W. Staples, Dr. Richard Criley, Clyde Imada, Frankie Sekiya, Jim Little, Leland Miyano, Richard Nii, Dr. and Mrs. Albert Yellin, Elsie Nakasone, Peter De Mello, Dale Sato, Rick Barboza, David Orr, and the staff of Honolulu Botanical Gardens, especially Olive Vanselow, Nathan Wong, and Josh Sand (for her habit and flower photos of *Cassia sieberiana*).

We are also indebted to Dr. David H. Lorence of the National Tropical Botanical Garden (NTBG) for his photo of the inflorescence of *Barnebydendron reidellii* and the habitat and flower photo of *Cordia lutea;* to Kenneth R. Wood of NTBG for his photo of *Brigamia insignis* in habitat; and to Steven P. Perlman, also of NTBG, for his habitat photo of *Brighamia rockii.* Thanks to Jonathan Liu for his habit photo of *Aloe vaombe* and to Donald R. Hodel for his foliage and bark photos of *Acacia seyal* var. *fistula,* the habitat photo of *Plumeria rubra* forma *acutifolia,* and his evapotranspiration paper. Thanks to Marian Rauch for her photo of flower color variations of *Kalanchoe blossfeldiana,* Susan Rauch Ferrell for her habit photo of *Thevetia thevetioides,* and Paul Weissich for photos of the inflorescences of *Heliotropium anamalum* var. *argenteum, Cassia roxburghii,* the flowers of

Hibiscus ovalifolius and *Bauhinia grandidieri,* habit photos of *Barnebydendron reidellii* and *Punica protopunica,* and the habit/bark photo of *Cassia seyal* var. *seyal.* We also acknowledge the coauthors of *Nā Lei Makamae*, Marie A. McDonald and Paul R. Weissich, for photos of *Plectranthus parviflorus* (habitat and foliage), *Lycium sandwicense* (habitat and foliage), *Bidens cosmoides* (flower), *Bidens menziesii* (habitat and flower), *Canavalia galeata* (habitat and flower), *Kokio drynarioides* (flower), *Sophora chrysophyllum* (habitat and flower). Thanks to Robert Webb for habitat photos of *Adenium somalense, A. somalense var. nova, A. arabicum,* and *A. obesum,* and to Brian Kemble for photos of habit and inflorescences of *Aloe dawei, A. sinkatana,* and habit photos of *Aloe greatheadii* and *Aloe nobilis.*

We give special recognition to Moriso Teraoka, who, with staff assistance from Kapiolani Community College, created a valuable xeriscape resource in his campus cactus and succulent garden. The collection of labeled species of water-conserving plants in the Koko Crater Botanical Garden (one of the five gardens in the Honolulu Botanical Gardens system) has been of inestimable value.

We are indebted to various members of the Hawai'i nursery industry for making available specimens for photo opportunities.

The authors especially acknowledge the support of the following in the successful effort to fund a prepublication subsidy to the University of Hawai'i Press in order to retain total text, quantity of color photos, and hardcover of the present volume: the City and County of Honolulu Board of Water Supply, the Friends of the Halawa Xeriscape Garden, Friends of Honolulu Botanical Gardens, the Landscape Industry Council of Hawai'i, American Society of Landscape Architects, Hawai'i Chapter, and Mr. Donald R. Hodel.

Special thanks are due to the Kaulunani Urban Forestry Program for funding assistance. Kaulunani is a program within the State of Hawai'i Department of Land and Natural Resources, Division of Forestry and Wildlife, an equal opportunity service provider and grantee of the U.S. Department of Agriculture Forest Service, Pacific Southwest Region.

Our very special gratitude goes to our editor, Nadine Little, of the University of Hawai'i Press. Her undying patience, tact, and professional guidance enabled us to produce a reader-friendly book. Mahalo.

KEY TO DESIGNATIONS AND ABBREVIATIONS

XERISCAPE ZONE DESIGNATIONS

Zone 1: smallest garden space for plants with the highest water use
Zone 2: second smallest garden space allocation for plants with low drought tolerance
Zone 3: moderate garden space allocation for plants with significant drought tolerance
Zone 4: largest garden space allocation for plants with low water use and high drought tolerance.

WEED RISK ASSESSMENT (WRA) DESIGNATIONS (HAWAII INVASIVE SPECIES COUNCIL)

Note: Only H(HPWRA) and EVALUATE designations are explicitly noted in the text. Species without explicit designations in the text have been assessed by the Hawaii Invasive Species Council and are not currently recognized as invasive in Hawaiʻi. See Appendix G for more information on the Weed Risk Assessment project.

H(HPWRA)—High Risk: species is likely to be invasive in Hawaiʻi or on other Pacific islands.

EVALUATE: no assessment of risk can be provided at this time because (1) important information is missing from the assessment or (2) the species possesses a combination of traits and characteristics that make its behavior difficult to assess using the WRA system.

ABBREVIATIONS

T	Thorns or spines
S	Species known to cause skin, eye, mouth, or throat irritation
P	Toxic sap, leaves, flowers, or fruit
+	Use special care in handling
N	Known nitrogen-fixing species
NP	Probable nitrogen-fixing species

GROUND COVERS AND SMALL SPECIES TO 2 FEET

Aeollanthus Repens

CREEPING AEOLLANTHUS

ZONE 4

Lamiaceae (Mint Family)

Spreading rather rapidly and rooting as it spreads, this tropical African plant will mound up to 12 inches in height. Its silvery gray foliage is a good color contrast for plants with dark green foliage. Its flowers are blue violet in color and are carried on vertical spikes several inches above the foliage. Give it a well-drained soil and full sun. It is wind, heat, and drought tolerant and might be tried in beach gardens. It is an excellent ground and bank cover to combat erosion.

AGAVE SPECIES

There are more than three hundred species of agave found in both North and South America. Almost all have narrow leaves tipped with a sharp spine. Listed are a select few of available species for the xeriscape.

Agave filifera

THREADY AGAVE T

ZONE 4

Agavaceae (Agave Family)

A small plant from Mexico, this species rather slowly develops a rosette of narrow leaves about 2 feet in diameter. Each leaf is edged with fine filaments providing an interesting textural accent. It is best in full sun in an open, well-drained soil. It is wind, heat, and drought tolerant. Plant it where its spines are out of reach of children.

Agave shawii

SHAW'S AGAVE T

ZONE 4

Agavaceae (Agave Family)

Another small agave, this species is native to southernmost coastal California. It forms an attractive rosette of leaves 2 feet long and double that in spread. It throws many basal offsets that slowly produce clumps. It is useful as a bank or ground cover or used as an accent in the border. Plant it where its spines are out of reach of children. It must have full sun and a well-drained soil. It is wind, heat, and drought tolerant. It is a slow grower.

Agave victoria-reginae

QUEEN VICTORIA AGAVE T

ZONE 4

Agavaceae (Agave Family)

This Mexican species is highly prized for its attractive leaves, which form a tight rosette 8 inches in height and a foot across. Each leaf is tipped with a sharp spine. Use it as a textural accent in the rockery or border in a place where its unique display may be appreciated close at hand. It does best in full sun in an open, well-drained soil. It is wind, heat, and drought tolerant. It is a slow grower.

ALOE SPECIES

The genus Aloe consists of about 360 species. They are native to widespread dry environments from South Africa north to tropical East Africa, as well as to Arabia and Madagascar, and there is one species from the Canary Islands. They vary from small rosettes of leaves an inch or so high to treelike species 50 feet in height. Most produce highly colorful, long-lasting inflorescences and are excellent xeriscape subjects, but they are, unfortunately, very underutilized in tropical gardens. Many species are, however, available from specialty commercial growers, mostly in California, and from local succulent hobbyists. Visit your local Cactus and Succulent Society shows, where collectors display their best specimens.

Nomenclature may be questionable. Aloes easily hybridize, but for the average home xeriscaper this should not be a deterrent. Many South African species will grow well in the tropics, but some from cooler areas fail. Local hobbyists are an excellent source of practical information. The tropical species of aloe thrive best. Flower arrangers value their beautiful flowers and even use entire plants in arrangements. The plants can later be returned to the garden and replanted. Aloes are moderate growers.

For common names, those marked with a single asterisk (*) are listed by the Institute for Aloe Studies; two asterisks (**) indicate names found online; and three asterisks (***) indicate that no common name was found by the authors.

Aloe aculeata

RED HOT POKER ALOE **

ZONE 4

Aloeaceae (Aloe Family)

This colorful aloe is native to northern South Africa. It will reach 2 feet in height and does not produce clumps. It has a tall, branched inflorescence growing well above the crown. Flowers, appearing in winter, may vary in color from yellow to orange or bicolors. It is a fine foliage and color accent in the garden, or use it as a border or in a mixed border in a well-drained soil in full sun. It is heat, wind, and drought tolerant.

Aloe brevifolia

SHORTLEAVED ALOE *

ZONE 4

Aloeaceae (Aloe Family)

A beautiful South African species, this aloe grows to 15 inches in height and forms dense clumps. Its flowers, carried on a stem 2 feet in height, are pale scarlet. It is heat, drought, and wind tolerant. Use it as a minor ground cover where you want an unusual color accent. Plant it singly or in groups in full sun in a well-drained soil. It is especially useful in the rockery but must be planted close at hand to appreciate its beauty. The figured flower close-up is actually a hybrid between *A. brevifolia* and an unknown species. Aloes readily hybridize.

Aloe deltoideodonta

TRIANGLE-TOOTHED ALOE **

ZONE 4

Aloeaceae (Aloe Family)

Native to Madagascar, this species grows to 9 inches in height, bearing several flower spikes up to 7 inches in height above the plant. It has scarlet flowers. It does not form a clump. As with other aloes, it presents a plant with strong tolerance to heat, drought, wind, and salt. Use it as a winter color accent or specimen in the rockery or border.

Aloe dorotheae

SUNSET ALOE *

ZONE 4

Aloeaceae (Aloe Family)

GROUND COVERS AND SMALL SPECIES TO 2 FEET

Forming broad-spreading clumps, this is a small species from Tanzania reaching only 10 inches in height. Under dry conditions, the entire plant turns red. It flowers abundantly, giving a strong orange- or red-colored accent to the xeriscape. Like other aloes, it is heat, wind, drought, and salt tolerant. Use it as a ground cover or bank cover for erosion control. It will thrive in any well-drained soil in full sun.

Aloe greatheadii

***

ZONE 4

Aloeaceae (Aloe Family)

A small, nonclumping aloe, this species from northern South Africa to Zimbabwe and east to Mozambique reaches 18 inches in height but develops a tall, branched inflorescence to 5 feet in height. Flowers, appearing during the winter months, are red. Use it in the rockery or in a mixed border as an accent. It must have well-drained soil and full sun. It is heat, wind, and drought tolerant.

Aloe haworthioides x *A. descoingsii*

***

ZONE 3

Aloeaceae (Aloe Family)

This is a popular hybrid between two diminutive Madagascan species. Although plants are only 2 to 3 inches in height, they form basal suckers creating rather dense clumps. Its delicate winter to spring inflorescence may reach 8 inches in height, bearing small blossoms in white or pale pink. It is heat and wind tolerant but only moderately drought tolerant. Use it in the dry garden in the rockery or as a foliar/textural accent in the border. Plant it in sun in a well-drained soil. It is a moderate grower.

Aloe hildebrandtii

***

ZONE 4

Aloeaceae (Aloe Family)

An unusual aloe, this shrubby species is multibranched at the base, slowly producing sprawling stems up to 3 feet in length and about 1 foot in height. Its flowers—red, orange, or yellow—cap a sparsely branched inflorescence about 1 foot in length. It is native to dry bush areas of Sudan and nearby areas of tropical East Africa and is heat, wind, and drought tolerant. Use it as a bank cover or in the rockery as an accent. It favors full sun and an open, well-drained soil.

Aloe juvenna

TIGER TOOTH ALOE **

ZONE 4

Aloeaceae (Aloe Family)

Growing slowly to about a foot in height, this rare species forms dense, spreading clumps bearing spikes of coral pink flowers. It is native to arid areas from Kenya, the Horn of Africa, and Socotra. Take care in handling, as the leaves are edged with rather sharp "teeth." Plant it in full sun in a well-drained soil. Use it as a foliar and color accent in the rockery, as an accent in a mixed border, or as a bank cover. It is heat, wind, and drought tolerant.

Aloe x *nobilis*

GOLDEN TOOTH ALOE **

ZONE 4

Aloeaceae

This is a South African hybrid possibly involving *A. mitriformis* or *A. distans* and *A. arborescens*. For the xeriscaper, however, it is only sufficient to note its garden value. It slowly develops rosettes of leaves 1 foot across and 1 foot high. These cluster to form a dense ground cover that can be used in a border or on a slope to deter erosion. Its orange to red flowers appear in late spring and summer. It is heat, wind, and drought tolerant.

Aloe rivae

***

ZONE 4

Aloeaceae (Aloe Family)

From arid areas of southern Ethiopia, Somalia, and northern Kenya, this species usually slowly develops a short, solitary stem crowned with leathery leaves up to 2 feet in length, producing a rosette 4 feet across. Its branched inflorescence stands another 2 feet above the rosette, carrying many flowers that are bright scarlet or yellow in color. It flowers in early fall and into spring. It is wind, heat, and drought tolerant and must have full sun and a well-drained soil. This provides a strong color accent in the xeriscape during a part of the year when other aloes are not yet in bloom. It can be used as a bank cover or as a specimen in the mixed border. In the literature, it is listed as a natural hybrid.

Aloe sinkatana

SUDANESE ALOE *

ZONE 4

Aloeaceae (Aloe Family)

Growing slowly to about 1.5 feet in height with a 3-foot spread, this aloe produces a sparsely branched inflorescence 3 feet in height tipped with flowers that may be scarlet, orange, or yellow. It is wind, heat, and highly drought tolerant, being native to an area in Sudan with an erratic rainfall averaging only 5 inches a year. Like other aloes, it makes a colorful accent, ground or bank cover, and looks good in a rockery. Give it full sun and an open, well-drained soil.

Aloe squarrosa

JEMAN ALOE **

ZONE 4

Aloeaceae (Aloe Family)

An unusual species from Socotra, this diminutive aloe grows to only 8 inches in height. It produces basal branches to form an attractive clump. It is described as an "undershrub," referring to its ability to grow in light shade. Flowers, held on a stem above the foliage, may be red or rose. It is heat, wind, and drought tolerant. Use it as a low ground or bank cover, in the rockery, or as a foliar accent in a mixed border. It is a moderate grower. Plant it in a well-drained soil.

Aloe striata

CORAL ALOE *

ZONE 4

Aloeaceae (Aloe Family)

subsp. *karasbergensis*

Considered to be one of the most attractive of all aloes, this South African decumbent species will reach 1.5 feet in height. Its grayish leaves are edged in pink and are "toothless." Its red flowers, appearing abundantly in winter, are held well above the leaf rosettes, providing a strong color accent. During the nonflowering period, the foliage itself is highly attractive and forms a strong foliar accent. Use this fine species in the mixed border, as a border itself, or as a bright accent in the rockery close at hand where its unusual qualities can be readily seen. It is a moderate grower and is heat, wind, and drought tolerant. *Aloe striata* subsp. *karasbergensis* displays more prominently striped leaves.

Aloe zebrine

TIGER ALOE **

ZONE 4

Aloeaceae (Aloe Family)

Forming clumps of leaf rosettes 2 feet across and 1 foot in height, this species is native to Angola and east to Mozambique. Leaf markings may be variable. The branched inflorescence is carried well above the foliage and may be reddish or brick orange red. It is heat, wind, and drought tolerant and may be used as a border winter color accent, a bank cover, or a ground cover. It is a moderate grower and does best in a light, well-drained soil.

Bryophyllum fedtschenkoi

FEDTSCHENKOI KALANCHOE

ZONE 4

Crassulaceae (Stonecrop Family)

'Marginata'

A low ground cover or bank cover plant rising to about 1 foot in height, this Madagascan species presents pink flowers above gray green foliage with a pinkish edge. A variant, 'Marginata', diplays foliage with strong markings. It is suited for use in the rockery and will cascade over a dark stone wall, making a striking accent. It is drought, heat, and wind tolerant.

Bulbine frutescens

YELLOW BULBINE

ZONE 4

Liliaceae (Lily Family)

'Hallmark'

'Hallmark'

Forming open rosettes of slender leaves up to 12 inches in length, this colorful, spreading succulent is native from low rainfall parts of the southern coast of South Africa north to Mozambique. The variety 'Hallmark' (orange-stalked bulbine) displays numerous orange flowers with yellow centers, while the species carries bright yellow blooms. They are produced well above the foliage and are seen almost every day of the year. It produces best in full, hot sun in a very well-drained soil. Once established it should be watered sparingly. Use it as a ground or bank cover, as a color accent, or displayed in a rockery composed of dark volcanic rock. In its native areas it is prized for its sap, which is used to heal burns and rashes.

Kalanchoe pumila

Bulbinella floribunda

BULBINELLA

ZONE 4

Liliaceae (Lily Family)

Native to South Africa, this tuberous rooted species will grow to 2 feet in height rather rapidly, forming clumps. Its flowers, appearing in early winter, are clustered in spikes of bright, clear yellow flowers. They are a good cut flower. Plant bulbinella in full sun in a well-drained soil. Use it as a strong color accent in a border or in the rockery.

Callisia fragrans

FRAGRANT CALLISIA

ZONE 4

Commelinaceae (Dayflower Family)

'Melnickoff'

'Melnickoff'

Native to southern Mexico, this useful species grows rather quickly to 18 inches, forming a dense ground cover, although under ideal conditions it may reach 4.5 feet in height. It fares well in both full sun and light shade and requires little watering once established. In full sun, foliage develops either a reddish leaf margin or a purplish flush. Foliage is uniformly green in shady situations. Its white flowers are fragrant. A variation is found in *C. callisia* 'Melnickoff', which produces variegated, more robust foliage. Both require good drainage and occasional pruning to produce a more visually attractive appearance. The cultivar can be used as an accent among dark rocks or as a dark green ground or bank cover. Both are successful in providing protection against erosion.

Callisia repens

INCH PLANT

ZONE 4

Commelinaceae (Dayflower Family)

Quickly forming a dense mat 2–3 inches in depth, this tropical American species makes an excellent ground cover. Use it also for an erosion control plant for slopes and in the rockery. It is drought, heat, and wind tolerant and thrives best in an open soil in full sun. The photos were taken of naturalized plants growing on a steep road cut, solid rock, and facing the afternoon sun in a dry location. It also thrives in moist areas.

Carpobrotus deliciosus

SWEET HOTTENTOT FIG

ZONE 4

Aizoaceae (Ice Plant Family)

This spreading, succulent species roots as it spreads, making it a prime selection for erosion control. It is native to South Africa, where it is found in areas ranging from coastal sand dunes to inland grasslands. Flowers are purplish, pink, or white. Its fruit is sweet and popular locally. It requires full sun and is heat, wind, salt, and drought tolerant. It is a good candidate for the beach garden and xeriscape. Provide excellent drainage.

Carpobrotus edulis

HOTTENTOT FIG, SOUR FIG

ZONE 4 H(HPWRA)

Aizoaceae (Ice Plant Family)

A sprawling succulent rooting as it spreads, this South African ice plant has naturalized in coastal California, where it is commonly seen as an erosion control species along highways. In Hawai'i, it prefers cool elevations but will succeed in beach areas in pure sand or an open soil. It will mound to about 12 inches. Flowers are pale yellow or cream. It is heat, wind, salt, and drought tolerant. Plant it in full sun.

Coreopsis grandiflora

COREOPSIS,
LARGE FLOWERED TICKSEED
ZONE 3
Asteraceae (Sunflower Family)

'Sunray'

Native from Michigan south to Florida, this species presents clumps of foliage 2 feet in height sporting an abundant supply of flowers throughout the year. The flower is a bright yellow, although many variations—including orange and mahogany red, as well as dwarf plants—have been developed by horticulturists. Double flowering cultivars such as 'Sunray' and 'Flying Saucer' are also available. Plant coreopsis in full sun in a well-drained soil. It is wind and heat tolerant and shows moderate tolerance to drought. In the xeriscape, use it as a colorful border, an accent, or in the rockery.

Coreopsis lanceolata

COREOPSIS,
LANCELEAF TICKSEED
ZONE 3
Asteraceae (Sunflower Family)

Another very similar sunflower, this species is native to roughly the same range as its cousin, *C. grandiflora,* but more adapted to higher elevations. Flowers are orange, although many cultivars have been developed showing double flowers of yellow and bicolors. Growing requirements and uses in the xeriscape are the same as for *C. grandiflora.* It will grow to 2 feet in height. Plant it in a well-drained soil in full sun. It is heat and wind tolerant and moderately tolerant of drought.

CRASSULA SPECIES

There are about three hundred species of *Crassula* native to Africa, mostly South Africa. They range in size from several inches to 10 feet in height. All are good choices for the xeriscape. We note but a few of those available.

Crassula capitella 'Camp Fire'

CAMP FIRE CRASSULA
ZONE 4
Crassulaceae (Stonecrop Family)

One of the smallest crassulas, this species reaches 3 inches in height but forms dense clumps. This cultivar is noted for the reddish infusion on leaf ends and sides, which makes it a good color accent. It also finds use as a ground cover or bank cover to prevent erosion. Flowers, appearing in spring, are white and not very conspicuous. Plant 'Camp Fire' in full sun for higher color, although it will also tolerate light shade. Give it a well-drained soil. It is wind, heat, and drought tolerant.

Crassula perforata

STRING OF BUTTONS
ZONE 4
Crassulaceae (Stonecrop Family)

This South African species forms a many-branched clump to 18 inches in height. Its flowers are a whitish yellow and are carried on slender stalks held 12 inches above the clump. It grows best in full sun but will tolerate light shade. Full sun may color foliage edges reddish. It must have good drainage. It tolerates heat, wind, and drought. Use it in the border or the rockery and as a foliage accent.

Crassula volkensii

VOLKEN CRASSULA

ZONE 4

Crassulaceae (Stonecrop Family)

Only 6 to 8 inches in height, this creeper from Kenya and Tanzania roots as it spreads, forming a dense mat. It makes an excellent ground cover, especially on slopes to prevent erosion. Or use it in the rockery. It is a moderate grower, producing best in full sun in a well-drained soil. It is wind, heat, and drought tolerant. A fine addition to the xeriscape.

Cyanotis somaliensis

SOMALI CREEPING DAYFLOWER

ZONE 4

Commelinaceae (Dayflower Family)

Forming a dense mat to 3 inches deep, this useful ground cover also provides effective erosion control. It roots down as it spreads. Its small flowers are a pale violet and are held well above the unusual foliage. Plants in cultivation are close to but possibly not conspecific with those in the wild that are found in Kenya, Somalia, and Yemen. In any event, it is a valuable plant for the dry garden. Provide a well-drained soil and a place in full sun. It is wind, heat, and drought tolerant.

ECHEVERIA SPECIES

About a hundred species of *Echeveria* are native from Texas to Argentina, mostly Mexico. They are referred to as "hens and chickens" due to the proliferation of basal offsets. We are able to cover only a few members of this useful genus.

Echeveria agavoides var. *prolifera*

MOULDED WAX

ZONE 4

Crassulaceae (Stonecrop Family)

Probably of Mexican origin, this useful plant slowly reaches 7 inches in height, its numerous offshoots forming dense clumps. Leaves are tinged with red. Use this species in full sun or light shade as a ground cover, bank cover to slow erosion, and as a color accent in the border or rockery. Give it a well-drained soil. It is wind, heat, and drought tolerant.

Echeveria carnicolor

CARNICOLOR HENS AND CHICKENS

ZONE 4

Crassulaceae (Stonecrop Family)

Another Mexican native, this species forms a tight rosette of pointed, grayish leaves tinged with purple. Flowers, carried on a slender stem several inches above the foliage, are salmon orange in color. Best results are had when planted in full sun in a well-drained soil. It is a moderate grower and tolerates wind, heat, and drought. Use it in a color border, as a ground or bank cover, or in the rockery. It is a color accent.

Echeveria colorata x *Echeveria elegans* 'Lendel Bedder'

LENDEL BEDDER HENS AND CHICKENS

ZONE 4

Crassulaceae (Stonecrop Family)

Hens and chickens are very popular garden plants, especially in California. Many hybrids have been produced. Many plants of unknown parentage are in the trade. This is a known cross between *E. colorata* and *E. elegans,* both Mexican species. It flowers from April through the next few months. Flowers are red and orange. Foliage is yellowish green to whitish, tinged with carmine above. *E. elegans* has lettuce green to pale glaucus green leaves. Flowers, appearing from March through the next several months, are rose and pale yellow orange. This combination has resulted in very tight, dense rosettes of foliage 6 to 7 inches in height, with many offsets. It is heat, wind, and drought tolerant. Plant this one in full sun in a well-drained soil. Use it liberally in the rockery or in a mortarless rock wall, as a ground cover, soil erosion bank cover, or as an accent among plants with dark foliage.

Echeveria 'Morning Light'

MORNING LIGHT HENS AND CHICKENS

ZONE 4

Crassulaceae (Stonecrop Family)

This beautiful cultivar is an example of many fine plants suitable for the dry garden but for which we can find no parentage. It is valued for its large leaves with highly colored margins and its light pink flowers. A good color accent, it is also useful in the border or for a ground or bank cover. Give it a place in full sun in an open, well-drained soil. It is wind, heat, and drought tolerant. It reaches about 6 inches in height.

Echeveria pulvinata

CHENILLE PLANT

ZONE 4

Crassulaceae (Stonecrop Family)

Native to Mexico, this useful species will grow to 8 inches in height and will slowly develop numerous branches. Its densely pubescent foliage, usually a rusty brown in color, makes this plant an interesting ground or bank cover. It does well in the rockery. Flowers are small, and red and yellow, and appear in winter. Give it a place in full sun in an open, well-drained soil. It is wind, heat, and drought tolerant.

Euphorbia decaryi

DWARF EUPHORBIA S

ZONE 4

Euphorbiaceae (Euphorbia Family)

A small Madagascan plant rather slowly reaching 6 inches in height but spreading to over a foot across, this species displays attractive chartreuse, yellow, or red flowers in spring. It is native to open dry forests and shrubland and is threatened by habitat destruction in the wild. It is heat, wind, and drought tolerant. Use it in the xeriscape as a color accent in a place where it is close at hand. Or use it in the rockery or as a ground cover in a limited space. It must have sun and an open, well-drained soil.

Evolvulus glomeratus subsp. *grandiflorus*

BLUE DAZE

ZONE 2

Convolvulaceae (Morning Glory Family)

Reaching 18 inches in height and spread, this Brazilian species forms a dense, compact mound of grayish green foliage carrying many bright blue flowers most of the year. It is a moderate grower. Give it full sun and plant it in a well-drained soil. Use it as a ground or bank cover, in the rockery, or as a border. It is a strong color accent. It shows moderate tolerance of drought.

Ipomoea cairica

Gaillardia x *grandiflora*

GAILLARDIA, BLANKET FLOWER

ZONE 3

Asteraceae (Sunflower Family)

'Goblin'

'Goblin'

'Fanfare'

'Fanfare'

Native to the central states, this American native is adaptable to the tropical xeriscape. It will grow to 2 feet in height, forming dense clumps. Many cultivars have been developed, showing flowers varying in color from orange with a yellow tip through burgundy red. There are semidwarf forms. All of them are highly colorful, tough, and salt tolerant and have been known to naturalize in onshore wind coastal beaches. It is moderately drought tolerant and highly tolerant of heat and wind and poor soils. It must have full sun, good drainage, and will not live long in wet situations. Gaillardias are an excellent color accent, border plant, or a subject for a difficult, hot spot in the xeriscape. Figured are 'Goblin' and 'Fanfare'.

GASTERIA SPECIES

There are about fifty species of *Gasteria* in the Lilia Aloeoceae (aloe family). All are native to South Africa, and all are useful in the xeriscape as small colorful accents in the border or in a rockery. Gasterias slowly form clumps of rosettes 6 inches or more in height bearing small but beautiful flowers on a stem held well above the clump. They thrive in a well-drained soil in full sun or very light shade. They are all wind, heat, and drought tolerant. The unidentified species pictured shows typical *Gasteria* form.

Gasteria sp.

MOTHER-IN-LAW'S TONGUE

ZONE 4

Aloeaceae (Aloe Family)

This South African species slowly forms a rosette of pointed, succulent leaves 6 inches in height. Suckers forming around the base gradually develop a clump. Flowers, colored pink with green stripes, are carried on slender stalks well above the clump. Gasterias are heat, wind, and drought tolerant. They require full sun and a well-drained soil. Use this diminutive plant as an accent, as a ground cover in a small space, or in the rockery. Plant it close to a path where its small size and color may be appreciated. Figured is an unidentified species called Mother-in-Law's Tongue. It forms clustered rosettes 6 inches in height. Flowers, appearing in winter, are carried on slender stems 18 inches in height.

Gazania rigens

PIED GAZANIA

ZONE 3

Asteraceae (Sunflower Family)

This South African species forms a rosette of leaves to 8 inches in height. It bears bright orange flowers banded with brown at the base of the petal, although horticulturists have developed a wide range of new colors from light yellow to deep red. Foliage varies from green to silvery. It performs best in full sun with well-drained soil in moderately higher, dry elevations. It is heat and wind tolerant and moderately tolerant of drought. It is a valuable color accent, ground cover, and bank cover and can be used to make a border planting.

Graptopetalum paraguayense

GHOST PLANT

ZONE 4

Crassulaceae (Stonecrop Family)

Despite its specific name, this is a Mexican succulent rather slowly reaching only 6 inches in height but spreading and rooting as it spreads. It makes a fine ground cover, especially where surface erosion is a problem, and is drought, heat, salt, and wind tolerant. Its unusual whitish lavender foliage offers a striking background for taller, dark-foliaged species. For optimal growth, give it a well-drained soil in full sun. Not figured is an interesting hybrid between this *Graptopetalum* and *Echeveria gigantea*. It is also useful in the dry garden.

HAWORTHIA **SPECIES**

There are about 160 species of haworthias, all native to South Africa, all useful in the xeriscape. We note only a few.

Haworthia aristata

BEARDED HAWORTHIA

ZONE 4

Aloeaceae (Aloe Family)

South Africa's eastern Cape of Good Hope is the home of this diminutive haworthia. It forms a rosette of leaves scarcely 4 inches across. Each leaf has translucent, fringed edges that make this a useful accent plant or, planted en masse, a ground and bank cover. Its flowers are white and somewhat inconspicuous, and are carried on a slender stem well above the plant. It is heat, wind, and drought tolerant and does best in full sun in an open, well-drained soil.

Haworthia attenuata

POINTED HAWORTHIA

ZONE 4

Aloeaceae (Aloe Family)

This small species forms rosettes 3 inches in height and double that in spread. These slowly produce many suckers, eventually making a dense mat of plants useful as a ground cover or bank cover to deter erosion. It is heat, wind, and drought tolerant. Give it a place in full sun or very light shade and a well-drained soil.

Haworthia coarctata

COMPACT HAWORTHIA

ZONE 4

Aloeaceae (Aloe Family)

Another mat-forming species, this haworthia reaches 8 inches in height and twice that in diameter and slowly produces many offsets. Use it in the rockery, in a mixed border, or as a ground cover or bank cover to reduce erosion. Give it a place in full sun or light shade in a well-drained soil. It is heat, wind, and drought tolerant.

Haworthia fasciata

ZEBRA HAWORTHIA

ZONE 4

Aloeaceae (Aloe Family)

A small, slow-growing species, the zebra haworthia reaches only about 1.5 inches in height, forming dense colonies that work well as bank covers, in the rockery, or in a low border. Full sun and a well-drained soil are required, although some growers report that a very light shade is tolerated. It is wind, heat, and drought tolerant.

Haworthia greenii

GREEN'S HAWORTHIA

ZONE 4

Aloeaceae (Aloe Family)

This species forms dense rosettes of speckled leaves about 8 inches in height and length. They become decumbent and slowly spread to make a fine ground cover or bank cover to control erosion. It is heat, wind, and drought tolerant.

Haworthia limifolia

NO COMMON NAME

ZONE 4

Aloeaceae (Aloe Family)

Native to Mozambique and south to Natal, this species rather slowly forms a rosette of pointed, succulent leaves 6 inches or more in length. Basal suckering eventually develops the plant into a dense clump. The nearly white flowers are borne on slender stems up to 1 foot in length. Under drought conditions, some references state that the foliage takes on a reddish color. This is another small accent plant for use in the rockery or border. It is heat, wind, and drought tolerant. Due to its popularity in traditional native medicine, this beautiful species has become highly endangered or extinct in its several once large wild populations. It is reported that in Durban, native medicinal markets sell approximately 479,000 *Haworthia limifolia* plants per year.

Heliotropium anomalum var. *argenteum*

HINAHINA KŪ KAHAKAI

ZONE 4

Boraginaceae (Borage Family)

Native to dry, coastal sites exposed to strong salt wind and spray, this diminutive, drought-tolerant, succulent Hawaiian species forms rosettes of silvery green leaves. Flowers are fragrant, varying from white to lavender. Both flowers and leaf rosettes are used to make a long-lasting lei. It is found throughout Hawai'i, growing in sand dunes, coral ledges, and rock in full sun and down to just above the reach of waves. It does not produce typical foliage inland. An excellent beach and xeriscape ground cover reaching 2 inches in height, it spreads readily to form a dense mat. It is also attractive as a potted plant for a sunny terrace at the beach.

Ipomoea cairica

KOALI

ZONE 3

Convolvulaceae (Morning Glory Family)

Koali, although a vine, is used most frequently as a ground cover so is here included. It is found throughout Hawai'i in both dry and wet areas and on dry beaches near sea level and up to 2,000 feet elevation. *Koali* is native to tropical Africa and Asia and may be a Polynesian introduction. It makes an excellent ground or bank cover a few inches deep but wide spreading. As figured, *koali* may also be used to cover an arbor or trellis. Plant it in full sun in any well-drained soil. In old Hawai'i, a lei of *koali* was worn during the ceremony and celebration marking the successful completion of a section of an irrigation system. It is a rapid grower.

Ipomoea pes-caprae subsp. *brasiliensis*

POHUEHUE, BEACH MORNING GLORY P

ZONE 2

Convolvulaceae (Morning Glory Family)

A useful vine, *pohuehue* is included among the ground covers, as it is exclusively used for that purpose. It is pantropical. It is found on beaches and sand dunes down to the water's edge but may also be grown inland. Although only a few inches in height, it is very wide spreading, rooting at the nodes. It is salt, heat, and wind tolerant but only moderately drought tolerant. It is a rather rapid grower.

Jaquemontia ovalifolia subsp. *sandwicensis*

PĀ'ŪOHI'IAKA

ZONE 4

Convolvulaceae (Morning Glory Family)

A diminutive, rapidly spreading endemic vine, this useful Hawaiian plant is native to sites close to the shore. It forms a mat about 2 inches in depth, rooting down at nodes, and it bears flowers all year varying from sky blue and blue to white. Foliage is green and leathery-succulent. There is one form with silvery gray foliage. It must have full sun and does well in sand, well-drained soil, and even on a rocky slope. It is very salt tolerant and should be used in beach gardens with onshore wind. It is also heat and drought tolerant. Although a vine, it is included in the ground cover section as this is its sole use in the garden.

KALANCHOE SPECIES

This succulent family includes about 125 species, mostly native to Africa and Madagascar. All are useful in the xeriscape. We list but an introductory few.

Kalanchoe blossfeldiana

MADAGASCAN CHRISTMAS KALANCHOE

ZONE 3

Crassulaceae (Stonecrop Family)

During the cool months, this species has become widely popular for its highly colorful, abundant flowers, which may be red, yellow, purple, or orange. In some areas a few flowers may appear during the year. Forming a clump up to 1 foot in height with an equal spread, this species is an important color accent in the dry garden used individually, en masse, or in a border. As the common name suggests, it is native to Madagascar. It is wind tolerant and moderately tolerant of heat and drought. It is a moderate grower.

Kalanchoe farinacea

FLOURY KALANCHOE

ZONE 4

Crassulaceae (Stonecrop Family)

Native to dry scrublands of Socotra, this attractive, whitish species is at home in the xeriscape. Growing to 1 foot in height with an equal spread, it works well as a ground cover, a bank cover for erosion control, or along a wall top or rockery where it will cascade. Its small but attractive flowers are red, carried on stems well above the plant, forming an interesting contrast with the foliage and providing a color accent. It is wind, heat, and drought tolerant and a moderate grower.

Kalanchoe luciae

PADDLE PLANT

ZONE 4

Crassulaceae (Stonecrop Family)

Somewhat paddle-shaped, succulent leaves are borne on a stem growing rather slowly up to 2 feet in height. Leaves develop an attractive reddish edge when plants enjoy full sun. Flowers, appearing in spring, are yellow and carried on a stem up to 3 feet in height. It is native to South Africa. Use it in the xeriscape as a color accent, in a border, or in the rockery. It is wind, heat, and drought tolerant. Provide a well-drained soil.

Kalanchoe marmorata

PENWIPER PLANT

ZONE 4

Crassulaceae (Stonecrop Family)

Growing to 12 inches in height, this species from Ethiopia, Somalia, and Kenya is an unusual foliage accent in the xeriscape. Its spots are enhanced when given full sun. It is a moderate grower, eventually developing several stems. Large, white tubular flowers on stems rising well above the foliage are seen in late spring and into summer. It does best in a well-drained soil. It is heat, wind, and drought tolerant. Use it as a textural accent among plain-foliaged plants or en masse for a ground cover.

Kalanchoe marnieriana

MARNIER KALANCHOE

ZONE 4

Crassulaceae (Stonecrop Family)

Forming a dense, low, spreading shrub to 12 inches in height, this Madagascan species roots as it spreads, providing a tough ground or bank cover for erosion control. Or it can be incorporated into a low, mixed border or in the rockery, where its large pink flowers can be appreciated. Flowers appear in winter. Plant it in full sun in a well-drained soil. It is wind, heat, and drought tolerant and a moderate grower.

Kalanchoe pumila

FLOWER DUST PLANT

ZONE 4

Crassulaceae (Stonecrop Family)

Growing to only 3 inches in height, this Madagascan ground cover spreads, rooting as it does. Its foliage is an unusual lavender gray in color. Flowers are lavender. It is heat, wind, and drought tolerant and can be a colorful accent in the xeriscape. Plant it in full sun in a well-drained soil. Use it in the rockery, in the border, as a bank cover for erosion control, or along a wall top. It will cascade. It is a moderate grower.

Kalanchoe synsepala

STOLON KALANCHOE

ZONE 4

Crassulaceae (Stonecrop Family)

Reaching as much as 2 feet in height, this Madagascan species is unique in its habit of producing stolons that protrude well beyond the mother plant. Each stolon develops a small plant at its tip that takes root as the stolon lengthens and bends to the ground. Flowers, held on a stem above the foliage, may be white, rose, or pale violet. Leaf margins are variable in shape from undulate to bluntly toothed. It is heat, wind, and drought tolerant and may be used as an accent specimen, a tall ground cover, or in a mixed border planting. It is a moderate grower.

Kalanchoe tomentosa

PANDA PLANT

ZONE 4

Crassulaceae (Stonecrop Family)

This Madagascan species displays velvety, gray green leaves edged with chocolate. It will reach 1.5 feet in height. Multiple plantings produce a good bank or ground cover or, used singly, a bright accent among dark rocks. Several foliage variations are available and all are attractive for the xeriscape. They are drought, heat, and wind tolerant. It is a rather slow grower and performs best in full sun when planted in a well-drained soil.

Aloe greatheadii

Melanthera integrifolia

NEHE

ZONE 4

Asteraceae (Sunflower Family)

An endemic succulent plant from hot, dry coastal areas of Hawai'i, Kure Atoll, and Laysan, *nehe* is a useful ground cover species, growing only a few inches in height but spreading. It roots as it spreads, forming a good erosion control plant. It is a rich green in color with yellow blossoms that can be used by the lei maker. It is heat, drought, wind, and salt tolerant and fares best in full sun in a well-drained soil. It is a moderate grower.

Melanthera lobata

NEHE

ZONE 4

Asteraceae (Sunflower Family)

Another *nehe* from dry coastal areas of Ni'ihau, O'ahu, and West Maui, this species forms a spreading ground cover up to a foot in height. Abundant flowers may be used in leis. It is heat, wind, drought, and salt tolerant. It is a moderate grower. A close relative, *Melanthera lavarum,* not figured, is frequently seen as a dense shrub 1–2 feet in height, but it can reach up to 6 feet in height. It is found in dry coastal habitats and open dry forests. Its bright yellow flowers are prized by the lei maker.

Pachyphytum compactum

THICK PLANT

ZONE 4

Crassulaceae (Stonecrop Family)

One of twelve species of pachyphytums, all of which are native to Mexico, this one will reach 12 inches in height. With abundant basal branching, it forms a dense ground cover, rooting down as it spreads. Use it also for erosion control on a bank. Flowers, appearing in the cool months, are conspicuously colored: the calyx is pink with bluish tips, while the petals are orange red with glaucous blue tips. Added to the fact that the foliage is glaucous and sometimes purplish, the entire plant becomes a strong color accent. It works well in the rockery. Plant it in full sun in a well-drained soil. It is wind, heat, and drought tolerant and is a moderate grower.

Plectranthus amboinicus

CUBAN OREGANO, SPANISH THYME

ZONE 3

Lamiaceae (Mint Family)

'Variegata'

'Variegata'

Possibly native to India, this spreading ground cover is now found throughout the tropics. It is prized for the aroma and flavor of its foliage, which is used in cooking. In India and Southeast Asia it has medicinal uses. Growing to 2 feet in height, it makes an excellent, deep ground cover. Plant it in full sun. It is heat and wind tolerant and moderately drought tolerant. *P. amboinicus* 'Variegata' carries green and white leaves, while an unnamed variation shows reddish leaf margins and another displays a light green "splotch" in the center of the leaf. All show small, whitish to pale lavender flowers. All are aromatic. This useful plant is a somewhat rapid grower.

Plectranthus parviflorus

'ALA'ALA WAI NUI, KUPALI'I

ZONE 4

Lamiaceae (Mint Family)

Native to dry areas of Australia to Malesia and to Polynesia, in Hawai'i it occurs on all the main islands except Kaho'olawe. It is variable in size, forming compact mounds only 3–4 inches in height in very dry sites to small shrubs 2 feet in height. It is succulent, with aromatic foliage. The small foliage from the low-growing variation is used in lei making. Plant it in full sun in a well-drained soil. It is a good, low ground cover and especially attractive when planted among dark lava rocks. It is heat, drought, and wind tolerant and moderately salt tolerant. It is a slow grower.

Plectranthus prostratus

PROSTRATE PLECTRANTHUS

ZONE 4

Lamiaceae (Mint Family)

Native to South Africa, this useful succulent grows to only 3–4 inches in height but produces 6-inch spikes of unique flowers a delicate mauve in color in late summer and early autumn. Rooting as it spreads, this species makes a fine ground cover or bank cover when planted in full sun in a well-drained soil. It is a moderate grower. It is wind, heat, and drought tolerant.

Portulaca grandiflora

MOSS ROSE, PORTULACA

ZONE 3

Portulacaceae (Purslane Family)

'Wildfire'

'Wildfire'

'Wildfire'

Growing to 6 inches in height but spreading to 18 inches across, this useful species is from northern South America. It has small, succulent leaves and flowers that may be 1 inch across. Flower color varies from rose and red to white and yellow, with bicolors and even stripes. There are also double-flowered forms. In the trade, a series of hybrids called "Wildfire Hybrids" is available. These may present improved colors and forms. It requires a light, well-drained soil and full sun. It makes an excellent, colorful ground cover in the xeriscape. It is heat, wind, and drought tolerant and is a fairly rapid grower.

Portulaca lutea

'IHI

ZONE 4

Portulacaceae (Purslane Family)

A widespread succulent plant from Polynesia, including Hawai'i, and from Micronesia and New Caledonia to Pitcairn Island, this species inhabits coastal sites close to the ocean. It is salt, wind, heat, and drought tolerant. Growing either as a prostrate plant or slowly reaching a few inches in height, *'ihi* makes a good beach garden and xeriscape ground cover. Plant it in full sun in an open soil with perfect drainage. It produces yellow flowers.

Portulaca molokiniensis

'IHI

ZONE 4

Portulacaceae (Purslane Family)

A rare, endemic succulent, this *'ihi* is found on exposed coastal sites close to the ocean on Molokini and Kaho'olawe Islands. Bearing clusters of lemon yellow flowers and slowly growing only to 8 or 9 inches in height, it is a good ground cover or color accent among lower growing seaside species. An excellent selection for the beach garden with strong onshore winds. It is salt, wind, heat and drought tolerant. It must have full sun and perfectly drained soil. Care must be taken to avoid overwatering.

Pseudognaphalium sandwicensium

'ENA'ENA

ZONE 4

Asteraceae (Sunflower Family)

Native to all the islands except Kaho'olawe, this attractive Hawaiian species is found on dry, coastal sand dunes and inland on dry cinders or lava. It reaches 5 inches in height, forming rosettes of its eye-catching, white silver leaves. Small whitish to pale yellow flowers are carried on slender, erect stems up to 18 inches tall. It is easily propagated from stem cuttings or seed, which is fortunate because the plant usually dies after flowering. *'Ena'ena* is heat, wind, salt, and drought tolerant, a good choice for the beach garden and xeriscape where it finds use as a ground cover, a bright accent among dark rocks, or in the border with other smaller but dark green–foliaged species. It is a slow grower.

Sansevieria trifasciata 'Hahnii'

BIRD'S NEST SANSEVIERIA

ZONE 3 H(HPWRA)

Agavaceae (Agave Family)

'Hahnii'

'Golden Hahnii'

'Moon Glow'

'Moon Glow'

A low-growing variety of the gold-banded sansevieria, this useful plant will slowly grow to 6 inches in height, forming a dense mat of its rosette-shaped foliage. It is robust, tolerating moderate drought, sun or light shade, wind, and moderate salt wind. Plant it in a well-drained soil and where small leaves will not fall from plants above and be "caught" in its bird's-nest shape. It is an excellent bank or ground cover for the xeriscape and can be used in arrangements. Among a number of cultivars, one, 'Golden Hahnii', displays bright yellow leaf margins, while another, 'Moon Glow', produces light gray green foliage.

Scaevola coriacea

DWARF NAUPAKA

ZONE 4

Goodeniaceae (Naupaka Family)

An endangered, endemic Hawaiian species, dwarf *naupaka* is found on coastal consolidated sand dunes on Maui and also from Kaupo, Maui, and two small islets off of Maui and Moloka'i. It is drought, wind, and salt tolerant. A slow grower, it is prostrate and attains only a few inches in height but is spreading, rooting as it spreads. Leaves are succulent. It makes a good ground cover in full sun. Give it a well-drained soil. It also thrives in sand.

SEDUM SPECIES

This is a large genus containing about six hundred species of succulents from the North Temperate Zone and mountains of the tropics. We present only three as an introduction to a useful group of plants for the xeriscape.

Sedum adolphii

GOLDEN SEDUM

ZONE 4

Crassulaceae (Stonecrop Family)

Native to arid places in Mexico, this succulent produces 2-inch rosettes of light golden tan leaves. Color is intensified during cool weather. Reaching 8 inches in height, stems branch, forming a dense clump. Abundant flowers are white and carried in clusters well above the foliage mass. This is an excellent ground cover or erosion control plant on banks, or a colorful addition to the mixed border. Provide a well-drained soil and full sun. It is heat, wind, and drought tolerant and is a moderate grower.

Sedum lucidum

SHINING SEDUM

ZONE 4

Crassulaceae (Stonecrop Family)

A Mexican species, this plant produces basal branches that develop into a dense clump up to 1.5 feet in height. Abundant white flowers are seen during the cool months. It is a fine ground cover and may also be used for erosion control on banks. It works well in the rockery. Plant it in full sun in a well-drained soil. It is wind, heat, and drought tolerant and is a moderate grower.

Sedum x *rubrotinctum*

CHRISTMAS CHEER

ZONE 4

Crassulaceae (Stonecrop Family)

Probably a garden hybrid, this plant derives its name from its bright red-tipped leaves. It grows to 8 inches in height and spreads widely, rooting as it spreads. Grow it in full sun in a well-drained soil. Use it in the border as a color accent, as a ground cover, or as a bank cover. It is a strong accent in the rockery. It is wind, heat, and drought tolerant and is a moderate grower.

***SENECIO* SPECIES**

This genus is one of the largest of flowering plants, having between two and three thousand species found in a wide range of environments, from cold temperate to tropical. Many are xerophytes. This publication provides but a small look into their wide potential in the dry tropical garden. Do not chew or ingest any part of any *Senecio,* as a few species are known to be toxic. See the suggested readings for more information.

Senecio mandraliscae

BLUE FINGER

ZONE 4

Asteraceae (Sunflower Family)

A colorful ground cover reaching 1 foot in height and spreading to 2 feet, this South African species is heat, wind, and drought tolerant. It thrives in full sun in a well-drained soil. It is prized for its bluish color and for its use as a ground or bank cover, a color accent in the border, or in the rockery. Some references indicate a possible toxicity. Avoid ingesting any part of the plant. It is a moderate grower.

Senecio vitalis

GROUNDSEL **P**

ZONE 4

Asteraceae (Sunflower Family)

Native to Namibia and South Africa, this colorful succulent grows rather slowly to 18 inches in height with a 20-inch spread. Its bluish, slender foliage makes it a good choice for the dry garden, where it provides a color and textural accent, or in the rockery. Flowers are yellow. It must have very good drainage and full sun. It is heat, wind, and drought tolerant. Take care not to ingest any part of a *Senecio* as some are thought to be toxic.

Sesuvium portulacastrum

SEA PURSLANE, ʻĀKULIKULI

ZONE 4

Aizoaceae (Ice Plant Family)

Found worldwide in tropical coastal areas just above the high water line, this useful succulent is heat, drought, wind, and salt tolerant. It forms a dense mat a few inches deep, rooting as it spreads. Foliage may be green, yellowish, or even red. Its small flowers may be almost white to lavender pink. Plant it in full sun in a well-drained soil. It is an excellent candidate for the xeriscape and the beach garden as a ground cover or as a bank cover for erosion control. It is a somewhat slow grower.

Sida fallax

ʻILIMA KŪ KAHAKAI

ZONE 4

Malvaceae (Hibiscus Family)

ʻ*Ilima* is found throughout the Hawaiian Islands as well as on other Pacific islands and west to China. It is highly variable relative to height, flower color, and leaf size and shape. In dry areas exposed to onshore salty winds, ʻ*ilima* is a prostrate ground cover only a few inches in height (ʻ*ilima kū kahakai*). At higher elevations it may be found as a shrub reaching several feet in height. Flower color ranges from light yellow to strong orange yellow to bronze red. These are treasured by the lei maker. ʻ*Ilima* is drought, heat, wind, and salt tolerant and fairly rapid in growth. The prostrate form makes a colorful ground color for both the beach garden and the xeriscape.

Stapelia gigantea

ZULU-GIANT

ZONE 4

Asclepiadaceae (Milkweed Family)

Growing to about 8 inches in height but spreading widely, this succulent is native to tropical and southern Africa and Mozambique. Its unusual flowers, 1 foot across, are sometimes called "carrion flowers" due to their odor. They are star shaped, fuzzy with long purplish hairs, and of a reddish brown and purplish brown color. It is drought, heat, salt, and wind tolerant and useful in the xeriscape as ground or bank cover or in the rockery. It is a moderate grower. Its unfortunate odor, which attracts flies that pollinate the flower, is undetectable from a few feet. There are many other stapelias and related succulent genera—such as *Huernia, Duvalia, Caralluma, Tavaresia, Echidnopsis,* and *Edithcolea*—that are useful in the xeriscape for ground and bank covers. Space does not permit their introduction here. *Stapelia gigantea* has become widely naturalized, at least in very dry places on Oʻahu.

Tagetes patula

FRENCH MARIGOLD

ZONE 2

Asteraceae (Sunflower Family)

Despite its common name, this annual species is native to Mexico and Guatemala. It grows to 18 inches in height, bearing an abundance of flowers variously colored yellow, golden, orange, orange brown to almost red, and bicolors. Both double and single flowers are available. It is a very popular bedding plant or border and color accent. Plant it in a well-drained soil in full sun. This species will not tolerate shade or wet sites. It is wind and heat tolerant and moderately drought tolerant. It is the source of an essential oil used in perfumery. The plant is currently being researched for suspected antifungal qualities. Deadheading will lengthen the period of flowering. The flowers attract butterflies. While not a shrub, it is included in the xeriscape palette due to its significant color. It is a rapid grower.

Talinum paniculatum 'Variegatum'

VARIEGATED TALINUM

ZONE 3

Portulacaceae (Moss Rose Family)

Native to tropical parts of Central and South America as well as the West Indies, this variegated ground cover will reach 2 feet in height. Its succulent, white and green foliage makes an attractive accent or ground cover. Although tolerant of considerable drought and full sun, variegated talinum will thrive under conditions of moderate moisture and light shade. Give it good drainage. It is wind tolerant. The species *T. paniculatum* is plain green and seen less frequently than its variegated form. Both are moderate growers.

Tetragonia tetragonioidies

NEW ZEALAND SPINACH

ZONE 3

Aizoaceae (Ice Plant Family)

Rapidly forming a mat a few inches deep and 4 feet across, this succulent herb is native to coastal sand dunes from Japan south to Australia and New Zealand and from temperate South America. In Hawai'i it has naturalized in similar environments. It is grown for its edible, spinach-flavored leaves, which may be eaten raw in salads or cooked. It must have full sun and excellent drainage. It shows good resistance to wind, heat, and salt and moderate tolerance of drought. Use it as a ground cover in the xeriscape or beach garden or in the edible garden combined with other moderately drought tolerant herbs.

Tetramolopium filiforme

TETRAMOLOPIUM

ZONE 4

Asteraceae (Sunflower Family)

This endemic and endangered Hawaiian species forms a dense, mounding little clump up to 8 inches in height, bearing small daisylike flowers, pale lavender in color. It is heat, wind, and drought tolerant. It is found on dry ridges and cliff faces on the leeward side of Mt. Ka'ala, O'ahu. It makes an attractive accent or ground cover in the rockery and in a mixed border. It will reseed itself. It is a slow grower.

Tradescantia pallida 'Purpurea'

PURPLE HEART, PURPLE TRADESCANTIA

ZONE 3

Commelinaceae (Dayflower Family)

'Purpurea'

'Purpurea'

A rapid-spreading Mexican species reaching 18 inches in height, this plant displays purple, succulent foliage and vertical sprays of small pink blossoms. A variant with pink-striped foliage is also available. It is wind and heat tolerant and moderately drought tolerant. It does best in full sun in a well-drained soil, providing a colorful bank or ground cover in the xeriscape.

Tradescantia sillimontana

WHITE VELVET

ZONE 4

Commelinaceae (Dayflower Family)

Native to northeast Mexico, this rapid-growing species forms a matlike ground cover several inches deep, rooting as it spreads. Use it for erosion control on a bank or as an accent among dark rocks. Its whitish foliage makes a good foil for darker species and for plants with colorful flowers or foliage. Plant it in full sun in a well-drained soil. It is heat, wind, and drought tolerant.

Tradescantia spathacea

OYSTER PLANT, MOSES IN THE CRADLE

ZONE 3

Commelinaceae (Dayflower Family)

'Variegata'

'Dwarf'

'Dwarf'

'Compacta Tricolor'

'Compacta Tricolor'

Reaching 2 feet in height, this tropical American succulent shows moderate tolerance to drought and is wind and heat tolerant. It is rapid growing given good, well-drained soil. It will grow in shade, but its color is enhanced by full sun. There are also varieties such as *T. spathacea* 'Variegata' with solid green and yellow stripes, as well as *T. spathacea* 'Tricolor' with green, pink, and yellow stripes, and *T. spathacea* 'Compacta Tricolor', a compact form. More commonly grown is its dwarf form listed as *T. spathacea* 'Dwarf'. It reaches 8 inches in height. All make good bank and ground covers, colorful borders, or potted specimens.

Tradescantia zebrine 'Discolor'

PURPLE WANDERING JEW

ZONE 3

Commelinaceae (Dayflower Family)

This useful Mexican species produces many runners, forming a mat up to 8 inches or more in depth. Runners root down as they grow, making this an excellent ground or cover to prevent erosion. It is heat, wind, and moderately drought tolerant and grows in almost any soil but best in a well-drained medium. It displays maximum color in full sun but will tolerate light shade. It is a fairly rapid grower. Use it as a color accent.

Verbena x *hybrida*

PERUVIAN VERBENA, COMMON VERBENA, GARDEN VERBENA

ZONE 3

Verbenaceae (Verbena Family)

Forming a spreading, dense mat to 6 inches deep, this tropical American perennial requires full sun and heat. It shows good drought tolerance. It roots at the nodes as it spreads, making it a good choice for a bank or ground cover. Brilliantly colored flowers in purple, blue, scarlet, or red appear most of the year. It will cascade over a wall top. Give this plant a well-drained soil. It is a fairly rapid grower.

Zinnia palmeri

BEACH ZINNIA

ZONE 4

Asteraceae (Sunflower Family)

Forming a compact shrub to 2 feet in height, this Mexican annual, previously thought to be *Zinnia maritima,* is prized for its many landscape qualities: it tolerates salt, wind, heat, and drought. It will thrive in beach sands or inland in any well-drained soil. It is rapid growing. Its small but abundant bright golden yellow flowers appear all year. They last a week as cut flowers and are valuable in lei *haku.* It reseeds itself. Use it as a ground cover, as a bank cover, in a color border, or as an accent among dark green or gray foliage plants. It thrives even in a 10-inch pot and provides a bright color accent for the full sun lanai or deck. It attracts butterflies.

SMALL SHRUBS
2–6 FEET

Abutilon menziesii

KO'OLOA'ULA

ZONE 3

Malvaceae (Hibiscus Family)

Endemic to Lana'i, east Maui, and Hawai'i, this rare, attractive shrub is prized for its use in lei making. Its pendant flowers are maroon grading to yellow in the center. It is wind and drought resistant and will grow to 5 feet in height. Use it as a filler between larger species with dark foliage, which will accent its light gray green leaves. Plant it in full sun. It must have a well-drained soil. Although an endangered species, it is readily available in Hawaiian plant nurseries. Position it so its flowers are available to the lei maker. It is a moderate grower.

Achyranthes splendens

HAWAIIAN ACHYRANTHES, 'EWA HINAHINA

ZONE 4

Amaranthaceae (Amaranth Family)

An endemic Hawaiian species, this silvery-leaved plant grows to 6 feet in height, making a bright accent in the xeriscape. Use it also as a low hedge behind shorter, dark green species or in the border. It is beautiful at night when uplighted. It must have full sun and a well-drained soil. It is drought, wind, and heat tolerant and shows moderate tolerance to salt air. A variety, *A. splendens* var. *rotundata* is similar but has more rounded leaves. Both are good choices for the xeriscape. Both are moderate growers.

Adenium swazicum

SUMMER IMPALA LILY **P**

ZONE 4

Apocynaceae (Dogbane Family)

Native to the east coast of South Africa, Swaziland, and north to Mozambique, this many-branched adenium will reach 6 feet in height. Its branches are supple, in some clones with a weeping habit, in others erect. Foliage is an attractive light gray green that enhances the color of its flowers, which may vary from rose, pink, and carmine to raspberry lavender. It is drought, heat, wind, and salt tolerant. Plant it in full, hot sun in a very well-drained soil. It is a moderate grower. Use it as a strong color accent. The weeping clones are especially valuable in a rockery or as a potted specimen on a sunny deck. Be wary of the sap, which is toxic.

Agapanthus praecox subsp. *orientalis*

AFRICAN LILY

ZONE 3 H(PWRA)

Liliaceae (Lily Family)

Reaching 4 feet in height, the African lily produces flowers in spring ranging from dark, purplish blue to blue and light blue and white. There are dwarf cultivars reaching only 12 inches in height. All produce good cut flowers and can be used to make a lei. It is heat tolerant, salt and wind tolerant, and moderately drought tolerant. Plant it in any well-drained soil, even sand, in full sun. It is a useful border or bedding plant. All are moderate growers.

Agave bracteosa

SPIDER AGAVE

ZONE 4

Agavaceae (Agave Family)

Growing among rocks and on cliffs in parts of Mexico, this agave produces a rosette of narrow green leaves up to 2.5 feet long that have neither marginal teeth nor spines at the tip. Its flower spike may reach 4 feet in height, bearing masses of showy white or pale yellow flowers. Plant it in full sun or light shade in a well-drained soil. It is heat, wind, and drought tolerant. Use the spider agave as a strong textural accent among rocks or in the mixed border. Unlike other agaves, it may be used where people contact is likely. It is a rather slow grower.

Agave gypsophila

GYPSUM CENTURY PLANT

ZONE 4

Agavaceae (Agave Family)

Growing to 3 feet in height, this small Mexican agave produces a flower spike up to 8 feet in height, carrying bright yellow blooms. Its grayish leaves are margined with soft spines, and the leaf tip bears a small spine. Plant it in sun or light shade in a well-drained soil. It is heat, wind, and drought tolerant. Use it among rocks, on a bank, or in a mixed border. It also makes a good container plant for a hot, windy lanai or deck. It is a slow grower.

Agave vivipara 'Marginata'

WHITE MARGINED AGAVE

ZONE 4

Agavaceae (Agave Family)

Native to Mexico and Central America, this distinctive agave bears leaves up to 4 feet in length bordered in white. In maturity the rosette of leaves may be up to 8 feet across, producing a branching flowering stem up to 15 feet in height, bearing abundant chartreuse flowers. Suckers are normally seen. This species is tolerant of heat, wind, and drought; use it as a strong accent in the rockery or broad border. Plant it where its sharp-tipped leaves cannot contact children or guests. It is a moderate grower doing best in an open, well-drained soil. Full sun is required for optimum growth.

Allamanda cathartica 'Thai Dwarf'

THAI DWARF ALLAMANDA **P**

ZONE 2 H(HPWRA)

Apocynaceae (Dogbane Family)

A useful cultivar growing to 3 feet in height, this plant is bushy, bearing light gray green foliage and small, bright yellow flowers. There is also a variety with medium green leaves. It must have full sun and a well-drained soil. Use it as a deep ground or bank cover, as a border or low hedge, or a single specimen providing a colorful accent. It is heat, wind, and moderately drought tolerant. Both are moderate growers.

Allamanda schotti

BUSH ALLAMANDA **P**

ZONE 2

Apocynaceae (Dogbane Family)

An evergreen shrub from Brazil, this yellow-flowered species will reach 5 feet in height. It flowers best in a fertile, well-drained soil in full sun. In the xeriscape, use it as a deep ground or bank cover, as a hedge, or in the border. It has moderate drought and salt tolerance and good wind tolerance. It is a moderate grower.

ALOE SPECIES

For common names, those marked with a single asterisk (*) are listed by the Institute for Aloe Studies; two asterisks (**) indicate names found online; and three asterisks (***) indicate that no common name was found by the authors.

Aloe broomii

SNAKE ALOE **

ZONE 4

Aloeaceae (Aloe Family)

Native to South Africa, this species reaches almost 5 feet in height, forming small clumps. Its flowers are yellow. It is heat, drought, salt, and wind tolerant and a good selection for the xeriscape. It may be used as a deep bank or ground cover, a seasonal color accent in the border, or in the rockery. As with other aloes, it fares best in a well-drained soil in full sun. It is a moderate grower.

Aloe debrana

***

ZONE 3

Aloeaceae (Aloe Family)

From fairly high elevations in Ethiopia, this unusual aloe grows to 3.5 feet in height, bearing a branched inflorescence twice that height carrying scarlet flowers. It prefers cool situations but will tolerate heat, wind, and drought. Give it an open, well-drained soil and full sun. It provides a color accent in the xericape and is usually grown as an individual specimen. It is a moderate grower.

Aloe cameronii

RED ALOE *

ZONE 4

Aloeaceae (Aloe Family)

A variable species, this aloe is native to Zimbabwe east to Mozambique. Usually growing to 4–5 feet in height, under ideal conditions it may grow to 6 feet in height, forming a solitary plant or small clumps bearing deep scarlet-colored flowers. It makes a strong color accent in the xeriscape or background planting for the smaller aloes or other succulents dark green in color. It is best in full sun in a well-drained soil and shows good tolerance of heat, drought, salt, and wind. It is a moderate grower.

Aloe camperi

NUBIAN ALOE **

ZONE 3

Aloeaceae (Aloe Family)

Reaching 3 feet in height, this species from fairly high elevations in Ethiopia does better in cool situations, but it will tolerate heat, wind, and drought. It develops a branched inflorescence twice that height, bearing orange to yellow flowers. It must have full sun and a well-drained soil. Use it in the rockery, as an accent in a mixed border, or on a bank for erosion control. It suckers from the base and may form groups up to 6 feet across. There are many useful hybrids between *Aloe camperi* and unknown species indicative of the ease of crossing that occurs. To be certain of obtaining only pure species from suppliers, ask for only hand-pollinated/controlled seedlings or offsets from known parentage. Both are moderate growers.

Aloe chabaudii

DWALA ALOE **

ZONE 4

Aloeaceae (Aloe Family)

From Zimbabwe, Zambia, and east to Mozambique, this colorful species forms dense clumps to almost 3 feet in height. Abundant coral pink, yellow to orange flowers are borne on branched inflorescences 3 feet in height. Full sun and an open, well-drained soil are basic to good growth, although one source indicates it will tolerate light shade. It may be used as a bank cover, as a brilliant winter color accent, or in a border. It is heat, wind, and drought tolerant. It is a moderate grower.

Aloe dawei

DAWE'S ALOE **

ZONE 3

Aloeaceae (Aloe Family)

Native to seasonally dry areas of central Africa and Kenya, this species shows only moderate tolerance of drought but is heat and wind tolerant. It will reach 6 feet in height, forming small clumps. Bright, reddish scarlet flowers are borne on branching stems 3 feet above the canopy. Use this species as a color accent, in the rockery, or to form a hedge. It must have full sun and a well-drained soil. It is a moderate grower.

Aloe decumbens

LANGEBERG RAMBLING ALOE **

ZONE 4

Aloeaceae (Aloe Family)

Another unusual "sprawler," this South African makes an excellent bank cover and erosion controller. Stems may reach 3 feet in length, forming a dense mat. It produces best in full sun in a well-drained soil and is tolerant of heat, drought, wind, and salt. Abundant flowers are scarlet. It is a moderate grower.

Aloe framesii

BITTER ALOE **

ZONE 4

Aloeaceae (Aloe Family)

Native to South Africa, this species is found on coastal sand dunes, where it slowly forms dense clumps of rosettes to 6 feet in diameter. The plant reaches a height of 2.5 feet, and its stems become decumbent and branch and rebranch. A fine choice for the xeriscape, it is heat, wind, drought, and salt tolerant. Its abundant scarlet flowers, sometimes tipped with light green, are seen in winter. It makes an excellent ground cover, bank cover, or color accent. Give it full sun and a well-drained soil. It is a moderate grower.

Aloe globuligemma

BUTTON ALOE **

ZONE 4

Aloeaceae (Aloe Family)

A highly attractive species from north Transvaal and Zambia to Mozambique, this aloe bears unusual multibranched inflorescences reaching to 3 feet in height above the foliage. The branches are held almost horizontally. Buds are dull red, opening to sulfur yellow to cream in color. Its rosettes of leaves may be 6 feet across and 3 feet in height. Producing creeping stems which root as they grow, this species forms dense clumps. It is heat, wind, and drought tolerant and may be salt air tolerant. Use it as a dense ground cover or erosion control bank cover. Plant it in a light, well-drained soil in full sun and where its color may be enjoyed. It is a moderate grower.

Aloe maculate

SOAP ALOE *

ZONE 4

Aloeaceae (Aloe Family)

A highly variable species, this aloe is native to a wide range of dry areas of South Africa. Growing to 3 feet in height, plants may be single stemmed or may form a cluster. Leaves bear elongated dull white spots in bands. Flowers, in distinctive flat-topped heads, are produced from spring to summer and may be variously colored from yellow to orange, red, and salmon pink. Spent flowering stems are persistent. Plant all aloes where access is easy and safe for removal of stems. The aloe figured is probably one of many available hybrids. Aloes hybridize readily.

Aloe mitriformis

GOLD TOOTH ALOE **

ZONE 4

Aloeaceae (Aloe Family)

Producing rosettes of foliage up to 2 feet across, this South African native branches freely from the base, developing a somewhat open, sprawling mat from 2 to 6 feet in depth. Flowers are dull scarlet. It is valuable in the xeriscape as a bank cover, where plants that are tolerant of heat, drought, salt, and wind are needed. It makes a good foliar and flower accent among dark rocks. It is best in full sun in a well-drained soil. It is a moderate grower.

Aloe mitriformis subsp. *distans*

***

ZONE 4

Aloeaceae (Aloe Family)

This is an unusual aloe in that it forms stems 3 to 6 feet long but sprawling, forming dense clumps. It carries plentiful flowers a dull scarlet in color. It is a very good bank cover for erosion control or used among rocks where it may fill low spots. Plant it in full sun in a well-drained soil. It is tolerant of heat, drought, wind, and salt. It is a moderate grower.

Aloe petricola

ZONE 4

STONE ALOE *

Aloeaceae (Aloe Family)

Flowering profusely during the winter months, this South African native will reach 3.5 feet in height. Usually seen as a solitary stem, it may form dense clumps several feet across. It must have full sun and a well-drained soil. Its flowers may vary from creamy white to yellow, orange, and red. Use it as a strong color accent, in the mixed border, as a bank cover, or as a low hedge. It is heat, wind, and drought tolerant and is a moderate grower.

Aloe x *salm-dyckiana*

SALM DYCK ALOE ••

ZONE 4

Aloeaceae (Aloe Family)

The parentage of this plant is questionable. It is no doubt a hybrid of South African background. Suffice it to say, however, for the landscaper it is enough to know that it will slowly reach 6 feet in height, forming dense, broad clumps. Its flowers are bright red and seen in winter. It makes a fine color accent and can be used as a bank cover, a hedge, screen, or windbreak. Plant it in full sun in a well-drained soil. It is heat, wind, and drought tolerant and is a moderate grower.

Aloe striatula

HARDY ALOE ••

ZONE 4

Aloeaceae (Aloe Family)

In xeric areas of South Africa, this species forms small clumps of attractive foliar rosettes up to 4 feet in height. Its flowers are a bright reddish orange to orange. It needs full sun in a well-drained soil to produce its best. As with other aloes, it is heat, drought, salt, and wind tolerant and finds good use in the xeriscape as a color accent or foliar accent among small-textured ground covers. It is a moderate grower.

Aloe suarezensis

•••

ZONE 4

Aloeaceae (Aloe Family)

A small aloe from dry parts of northern Madagascar, this species produces a single trunk attaining 2.5 feet in height. A branching inflorescence rising another 2 feet above the trunk displays numerous scarlet flowers. Give it full sun, well-drained soil, and use it as a color accent or in a mixed border. It is heat, wind, and drought tolerant. It is a moderate grower.

Aloe vera

MEDICINAL ALOE •, BARBADOS ALOE, PANINI ʻAWAʻAWA

ZONE 4

Aloeaceae (Aloe Family)

var. *chinensis*

Forming a broad, dense clump 2 feet in height, this aloe produces attractive yellow flowers on 3-foot stalks in winter or spring. Its actual origin is unknown, but it is possibly from the Mediterranean region or the Arabian Peninsula. It is of ancient use due to its medicinal properties and was noted in the first century AD by the Romans and much earlier by the Assyrian-Babylonians two thousand years ago. It was used by early Egyptians. *Aloe vera* is now grown throughout the world and is a valued plant in local kitchen gardens for its sap, which has the ability to reduce the pain of burns and cuts and insect bites. It effectively greatly reduces the pain of sunburn and has been used experimentally to counteract radiation burns. *Aloe vera* is now a common ingredient of many over-the-counter medications. It is drought, heat, wind, and salt tolerant. It prefers full sun in any well-drained soil. Use it as a bank cover, in the rockery, as a border, and as a tubbed specimen for a hot, windy lanai. An orange-flowering variation found in the trade is referred to as *Aloe vera* var. *chinensis,* a name of questionable validity. It has the same needs and uses as the species.

Anisacanthus thurberi

DESERT HONEYSUCKLE

ZONE 3

Acanthaceae (Acanth Family)

From dry places in the southern United States, this evergreen shrub grows to 5 feet in height. Orange flowers are borne much of the year. It requires full sun and a well-drained soil. It has good drought and wind tolerance and moderate salt tolerance. In the xeriscape it can be used as a tall bank or ground cover or as a colorful background. It is a moderate grower.

Argemone glauca

PUAKALA T

ZONE 4

Papaveraceae (Poppy Family)

A Hawaiian endemic species, *puakala* is found in dry, leeward areas from sea level up to 3,000 feet. It is highly drought, heat, and wind tolerant and will tolerate a moderate amount of salt. It will reach 2–3 feet in height, bearing large, crepey, white flowers. Its leaves are a bright bluish silver color and are prickly. It must have full sun and good drainage. It usually reseeds itself but is not invasive. Use it as a bright accent among rocks or as a bedding plant above a low, dark green ground cover. Its sap and seeds were traditionally used by Hawaiians to ameliorate toothache. It is a rather slow grower.

Argemone mexicana

MEXICAN POPPY T

ZONE 4

Papaveraceae (Poppy Family)

Dry places in the West Indies and Mexico are the home of this annual poppy that grows to 2 feet in height. Flowers are plentiful and a bright yellow in color. It is heat, wind, and drought tolerant. Easily grown in full sun in a well-drained soil, the Mexican poppy provides a good color accent in the xeriscape. It is a rapid grower.

Artemisia australis

ʻĀHINAHINA, HINAHINA

ZONE 3

Asteraceae (Sunflower Family)

This endemic shrub bears aromatic, whitish silver foliage. Leaf size and shape variability is under study. It reaches up to 18 inches in height with an equal spread and is found on exposed, moist windward cliff faces on all the main islands. It displays moderate tolerance of drought and great tolerance to wind. It makes a fine color accent and can also be used as a ground cover or bank cover. Provide a well-drained soil and plant it in full sun. It is a rather slow grower.

Artemisia mauiensis

'ĀHINAHINA, HINAHINA

ZONE 3

Asteraceae (Sunflower Family)

Although endemic to the arid upper slopes and crater of Haleakalā, this species performs satisfactorily at low elevations. It will reach 3 feet in height with a similar spread if given full sun and perfect drainage. It will not tolerate soggy soil. Use it as a bushy ground or bank cover. Its fine, silvery foliage makes it a dramatic accent when planted over a low, dark green ground cover or in the rockery. It is drought, wind, and heat tolerant and is a moderate grower.

Asparagus densiflorus 'Meyers'

MEYERS ASPARAGUS, FOXTAIL ASPARAGUS

ZONE 2 (HPWRA)

Liliaceae (Lily Family)

A South African evergreen shrub forming clumps reaching 2 feet in height, this tough plant shows moderate tolerance of drought and salt. It will grow readily in any well-drained soil. In the xeriscape it is useful in the rockery, as a bank cover, a container plant, or as an accent over lower, dark green ground covers. Plant it in full sun or light shade. Removal of its bright red seeds before maturity will prevent its becoming invasive. Pull seedlings. Entire stems are used in arrangements, while pieces of stem may be incorporated in a lei. It is a moderate grower.

Asparagus densiflorus 'Sprengeri'

SPRENGER ASPARAGUS

ZONE 2 H(HPWRA)

Liliaceae (Lily Family)

This South African species forms a clump of arching stems reaching 6 feet in length, producing a bright green shrub reaching 4 feet in height. It does well in any well-drained soil in full sun or light shade. It has good drought and salt tolerance. Small white flowers are followed by bright red fruit, which should be removed prior to maturity to prevent invasiveness. Pull any seedlings for the same reason. It makes a useful bank cover or, planted along the top of a retaining wall, an attractive cascade of foliage. It can also be used in a hanging basket or window box. Stems may be used in making a lei or large arrangements.

Aspidistra elatior

CAST IRON PLANT

ZONE 2

Liliaceae (Lily Family)

'Variegata'

'Milky Way'

An extremely tenacious Chinese species, the cast iron plant is prized for its ability to withstand abuse, be it deepest shade or, once established, lack of moisture. It shows moderate drought, heat, and wind tolerance. Its dark green leaves, up to 2.5 feet in length, rise from underground stems, forming a dense ground cover. There is a variety with white-banded leaves, *A. elatior* 'Variegata', as well as a cultivar named 'Milky Way' that shows spotted foliage. Plant the cast iron plant in light shade in a good, well-drained soil. Use it en masse in a problem area with low light or as a house plant. Or use it as an accent grouping above a light green, small-textured ground cover. Although capable of tolerating considerable drought, it produces a visually better specimen given a modicum of water. It is a slow grower.

Bidens hillebrandiana
(see next chapter)

Bidens mauiensis
(see next chapter)

Bougainvillea 'Temple Fire'

TEMPLE FIRE BOUGAINVILLEA

ZONE 3

Nyctaginaceae (Four O'Clock Family)

'Temple Fire'

'Temple Fire'

'Menehune'

Used as a shrubby ground or bank cover or color accent, this cultivar will reach 4 feet in height. It shows moderate tolerance to drought and salt with good tolerance to heat and wind. A similar dwarf bougainvillea is *B.* 'Menehune', which produces pale purple bracts. Both make a good border or low hedge. Both are moderate growers.

Bryophyllum gastronis-bonnieri

BONNIERI BRYOPHYLLUM

ZONE 4

Crassulaceae (Stonecrop Family)

This species is native to Madagascar. It will reach up to 6 feet in height, bearing large, succulent, "folded" leaves up to 20 inches in length. The leaves produce a plantlet at each tip that can be used to propagate more plants. The inflorescence is quite spectacular, appearing in winter and showing abundant, large, pendant flowers of yellow red. It is heat, wind, and drought tolerant. It must have full sun and a well-drained soil. Use this fine plant as a strong color accent or as a ground or bank cover planted en masse. Whole plants may be carefully dug and used in arrangements and later replanted in the xeriscape.

Calandrinia grandiflora

ROCK PURSLANE

ZONE 3

Portulacaceae (Purslane Family)

Bearing succulent, gray green leaves up to 8 inches in length, this tough, spreading ground cover from warm, dry parts of Chile may reach 3 feet in height. Its purplish flowers last only one day but are replaced by new flowers daily. It is almost ever blooming. Rock purslane thrives in areas exposed to full, hot sun and in well-drained soil. It will not tolerate "wet feet." It provides a fine ground cover and soil erosion controller, while sparkling its bright flowers against the thick foliage. It is wind, drought, and heat tolerant and probably fairly salt tolerant. It is a slow grower.

Callistemon viminalis 'Little John'

DWARF BOTTLEBRUSH

ZONE 2

Myrtaceae (Eucalyptus Family)

One of many variations of the weeping bottlebrush, this Australian selection brings its bright red flowers and bluish, gray green foliage to accent the xeriscape during spring months. Growing somewhat slowly, it will reach 3 feet in height with a canopy spreading to 5 feet. Once well established, it will tolerate moderate drought, heat, and wind. Plant it in full sun in good, well-drained soil. Use it as a low hedge or border.

Capparis sandwichiana

MAIAPILO, HAWAIIAN CAPER

ZONE 4

Capparaceae (Caper Family)

Usually seen as a decumbent shrub reaching 3 feet in height in hot, dry coastal areas, occasionally an upright specimen may be seen reaching 6 or more feet in height. It is heat, wind, drought, and salt tolerant. Leaves are an attractive gray green and are fleshy. Its large white flowers bloom at night, turning pink with age. Fruits are decorative and orange in color but with a fetid odor close at hand. This is a colorful ground cover among dark lava rocks or over a heavy, black cinder mulch. It may be pruned to a shrub form. Plant it in full sun in a very well-drained soil. *Maiapilo* also occurs on Midway Atoll, Pearl and Hermes Atoll, and Laysan. It is a slow grower.

Carissa macrocarpa 'Prostrata'

PROSTRATE CARISSA

ZONE 3

Apocynaceae (Dogbane Family)

'Prostrata'

'Prostrata'

'Variegata'

'Variegata'

Small-leaved, dark green, dense foliage is an attractive quality of this prostrate form of the common carissa. It grows to 2 feet in height, forming a dense ground cover dramatizing light-foliaged plants above, a bank cover, or in the rockery. Two similar cultivars, 'Tutleyi' and 'Tomlinson', reaching 2 feet in height, provide similar uses in the xeriscape. Other carissa dwarfs include 'Green Carpet' and 'Boxwood Beauty', which are prostrate, growing to 3 feet in depth. They occasionally throw a vertical shoot that must be removed. All make excellent ground covers or bank covers. There are also variegated forms. These all have good heat, drought, and wind tolerance and considerable tolerance to salt air. They are happy in any well-drained soil in full sun and are moderate growers.

Catharanthus roseus

MADAGASCAR PERIWINKLE **P**

ZONE 3

Apocynaceae (Dogbane Family)

This Madagascan shrub will reach 2 feet in height. The species bears dull, rose-colored flowers all year. Modern hybrids, however, have been developed ranging in color from pure white, red, and rose pink to purple and are very popular as bedding plants, ground covers, or borders. It provides good color in the xeriscape all year. It displays good heat, wind, drought, and salt tolerance. Its sap has been medically exploited to combat certain kinds of cancer. It is a rapid grower.

Colubrina asiatica

ʻĀNAPANAPA

ZONE 3

Rhamnaceae (Buckthorn Family)

This native species is found as a strand plant in areas close to the ocean on all the major islands except Maui and Hawaiʻi. It is widely distributed in coastal areas from East Africa east through the Indian Ocean to Southeast Asia, Malesia, Australia, and the Pacific Islands. It is a moderate grower, usually seen as a dense, somewhat scandent shrub reaching 6 feet in height. References indicate that it occasionally reaches small tree size. Its foliage is a dark, shiny green, making it a valuable plant in the beach garden and xeriscape, where foliage is frequently grayish or light green. Use it as a high ground or bank cover or as a low hedge or screen. It is heat, drought, wind, and salt tolerant. It prefers full, hot sun but will also grow satisfactorily in light shade. Flowers are inconspicuous.

Crassula ovata

ZONE 4

LARGE JADE PLANT, JADE PLANT, JADE TREE

Crassulaceae (Stonecrop Family)

'Grey Bark'

'Tricolor'

'Compacta'

Although reaching 12 feet in height in its native South Africa, the large jade plant usually is seen growing to 6 feet in height in local gardens. It grows rather slowly, developing a dense shrub that is tolerant of wind, heat, drought, and salt. Its foliage is succulent. Many forms have been developed, including those with elongated foliage and foliage of a blue green or reddish color. Best coloration is seen when growing these plants in full sun. Figured are the following cultivars: 'Grey Bark', 'Pink Puff', and 'Tricolor'. Another, *Crassula ovata* 'Compacta', also known as 'Nana', grows only to 3 feet in height, carrying small, rounded leaves. Use the "jades" for low borders or hedges or as color accents in the xeriscape. Smaller leaved forms have been used by the lei maker. Flowering is more abundant at cooler elevations. There are many crassulas and members of the Crassulaceae available to the xeriscaper. Space limitation does not permit their inclusion here.

Crassula perfoliata var. *falcata*

FALCATE CRASSULA, AIRPLANE PLANT

ZONE 3

Crassulaceae (Stonecrop Family)

Reaching 2.5 feet in height, this South African succulent provides color accents from both its red flowers and gray green leaves. It is a rather slow grower but eventually forms dense clumps, making it a good ground cover and bank cover. It does well in a rockery. Give it a place in full sun and in a well-drained soil. It is wind, heat, and drought tolerant.

Crinum asiaticum

SPIDER LILY, GIANT LILY, POISON BULB **P**

ZONE 2

Liliaceae (Lily Family)

'Variegatum'

An evergreen plant from Southeast Asia, this species forms a dense clump reaching 5 feet in height. It produces large clusters of fragrant flowers much of the year. There are forms with foliage and fruit tinged with purple and one with leaves striped with white. Spider lily is at its best in full sun in a well-drained soil and is heat, wind, and salt tolerant but only moderately drought tolerant. Flowers, foliage, and fruit are used in arrangements. It is useful in the xeriscape as an accent, where its rosette of large leaves is contrasted with a fine-foliaged, dark green ground cover. It is useful in the border or in a mass planting. Both flowers and fruit are used in arrangements. The fruit is toxic and should be kept away from children.

Cuphea hybrids

CIGAR FLOWER, FIRECRAKER PLANT, PUAKĪKĀ

ZONE 2

Lythraceae (Crape Myrtle Family)

This is an evergreen Mexican shrub reaching 3 feet in height. It is rather slow growing and spreading. It is best when given full sun and a good, well-drained soil. It is moderately drought and wind tolerant but shows no tolerance to salt. In the xeriscape it makes a good shrubby ground or bank cover or bedding plant, and a very good color accent. Several color forms are available. Plant it where its colorful flowers are accessible to the lei maker.

Dasylirion wheeleri

SPOON FLOWER, SOTOL

ZONE 4

Agavaceae (Agave Family)

Very slowly producing a trunk to 3 feet in height, bearing a spherical crown of slender leaves 3 feet in length, this species is a real xerophyte, tolerating drought, heat, and wind. It is native to dry areas of Mexico and western Texas to Arizona. It provides an excellent color and foliage accent in the xeriscape. It is useful in the night garden with uplighting. In its native areas, the young, bluish gray leaves are used to make heavy, durable baskets. Flowers are greenish white and are borne on a slender spike 9 to 15 feet in height.

Euphorbia milii

CROWN OF THORNS **T**

ZONE 4

Euphorbiaceae (Euphorbia Family)

A drought-tolerant plant from Madagascar, this popular and colorful species reaches 4 feet in height. It has been the subject of much breeding and cultivar selection, providing choices of dwarf forms and taller forms. The species bracts are red; new hybrids provide bract colors ranging from white to chartreuse, pink, coral, and red. It requires full sun in any well-drained soil. It is heat, wind, salt, and drought tolerant. Care should be taken when handling. This species and its cultivars and hybrids are among the most colorful subjects for the xeriscape. They can be used as low hedges or borders or as solid bedding plants. Plant so adequate maintenance space is provided to prevent contact with its thorns. It is a rather slow grower.

Euryops chrysanthemoides x *E. speciosissimus*

EURYOPS

ZONE 3

Asteraceae (Sunflower Family)

Euryops chrysanthemoides x *E. speciosissimus*

Euryops pectinatus subsp. *pectinatus*

Producing its bright yellow flowers most of the year, this evergreen hybrid South African shrub will reach 3–5 feet in height. It grows rapidly in full sun in a very well-drained soil. It has good drought and salt tolerance. Use it as a low hedge, in the border, or as background for lower flowering specimens. Heavy pruning produces branching, resulting in significantly more flowering. Another similar South African species, *Euryops pectinatus* subsp. *pectinatus*, the fuzzy euryops, with silvery gray foliage, provides similar landscape value and bears lemon yellow flowers. Care and horticultural requirements are the same as above.

Gardenia augusta 'Radicans'

TRAILING GARDENIA, PROSTRATE GARDENIA

ZONE 2

Rubiaceae (Coffee Family)

A dense, evergreen cultivar, this handsome plant will reach 2 feet in height with a spread of 3 feet. Its white blossoms are fragrant. Performing best at cool elevations in light shade in an acid, well-drained soil, it shows moderate drought and wind tolerance but has no salt tolerance. It makes a fine dark green ground cover. Use it where its fragrance may be enjoyed. It grows rather slowly.

Gossypium tomentosum

MAʻO, HAWAIIAN COTTON

ZONE 4

Malvaceae (Hibiscus Family)

A Hawaiian endemic shrub growing to 6 feet in height, this useful plant is found in hot, arid, coastal areas on all the islands except Hawaiʻi. Its gray green foliage is accented by bright yellow flowers seen most of the year. They are used by the lei maker. Plant it in full sun. Use *maʻo* as a bank cover, deep ground cover, or as a light foil for larger species with dark foliage. *Maʻo* is heat, wind, drought, and salt tolerant and is a moderate grower.

Hibiscus ovalifolius

ROCK'S HIBISCUS

ZONE 4

Malvaceae (Hibiscus Family)

Collected by Dr. Joseph Rock on Kauaʻi in 1913, this tough hibiscus was known as *Hibiscus rockii* until recent research indicated its origin as tropical Africa to South Africa, Madagascar, and the Mascarenes. In any event, it is a useful plant for the xeriscape. Slowly growing up to 4 feet in height, it spreads to form an evergreen, shrubby cover that produces large yellow blossoms most of the year. It is drought, heat, and wind tolerant and moderately tolerant of salt. Plant it in full sun in a good, well-drained soil. Use it as a colorful ground or bank cover or a low hedge.

Rosmarinus officinalis

Ipomoea carnea subsp. *fistulosa*

BUSH MORNING GLORY

ZONE 3

Convolvulaceae
(Morning Glory Family)

Possibly growing as tall as 10 feet, this colorful Brazilian species should be kept heavily pruned back to several feet in height to prevent its characteristic and unattractive legginess. It is a moderately rapid grower. Clusters of bright rosy pink flowers appear at branch ends much of the year. It will thrive in almost any well-drained soil, withstanding drought. Some protection from wind will prevent damage to its large blooms. It is not salt tolerant.

Jasminum sambac

ARABIAN JASMINE, PĪKAKE

ZONE 3

Oleaceae (Olive Family)

Originating in India, this rather sprawling shrub will reach 5 feet in height. Planted in full sun in a well-drained soil, it will produce a profusion of white, fragrant flowers during the warm months of the year. There are large-flowered and double-flowered forms available. Use it for a border, filler, or tall ground cover but plant it where its blossoms may be readily harvested. It is one of the most popular lei flowers. It shows moderate tolerance of drought, heat, wind, and salt and is a moderate grower.

Jatropha podagrica

GOUT STALK **P +**

ZONE 4

Euphorbiaceae (Euphorbia Family)

Usually seen as a potted specimen, gout stalk, native to Central America, will reach 5 feet or more in height when planted out in the xeriscape. It is heat, wind, and drought tolerant. Seeds are toxic. Plant this attractive plant in full sun in a very well-drained soil. It works best as a single specimen accent and is particularly handsome among dark rocks or above a dark green, low ground cover. It is rather slow growing.

Juniperus procumbens

JAPANESE GARDEN JUNIPER, SHORE JUNIPER

ZONE 3

Cupressaceae (Cypress Family)

This evergreen Japanese shrub reaches 2 feet in height with a spread of 6 feet. Plant it in full sun. It requires a well-drained soil. It displays moderate drought and salt tolerance. Use it as a ground cover, as a bank cover, or in the rockery. It will cascade over a wall top. Several cultivars are available. It is a rather slow grower.

Justicia brandegeana

SHRIMP PLANT

ZONE 3

Acanthaceae (Acanth Family)

Native to Mexico, this species will reach 6 feet in height. Planted in full sun in a good, well-drained soil, it will produce its unusual flowers all year. There is a form with yellow flowers and another with dark red blooms. It is partially drought tolerant and heat and wind tolerant. Use it to form a border or hedge or as a facer for taller plants. It is a moderate grower.

Kalanchoe longiflora

LONG FLOWER KALANCHOE

ZONE 4

Crassulaceae (Stonecrop Family)

A colorful ground or bank cover, this succulent from Natal in South Africa will slowly produce a compact mound to 2.5 feet in height with an equal spread, its stems rooting as it expands. It is useful in the rockery or in a mixed border. Its flowers are yellow but not conspicuous. To produce best growth and color, plant it in full sun in a well-drained soil. It is wind, heat, and drought tolerant.

Kalanchoe orygalis

ZONE 4

STICKY FINGER, COPPER SPOONS

Crassulaceae (Stonecrop Family)

Reaching 4 feet in height with an equal spread, this colorful Madagascan xerophyte is heat, wind, and drought tolerant. It is a rather slow grower. Plant it in full sun in a well-drained soil. It makes an interesting color accent among gray or silvery foliage and can be used to make a low hedge or border or in the rockery.

Lantana camara modern hybrids and cultivars

LANTANA **P T**

ZONE 4

Verbenaceae (Verbena Family)

The wild type, or species lantana, is considered one of Hawai‘i's worst invasive species, readily naturalizing in hot, dry areas statewide, thereby destroying valuable pasture and grazing lands. Biological controls, however, have gone a long way toward almost eliminating these populations. Modern cultivars and hybrids appear not to present an invasive problem. They rarely fruit, and the biological controls are still at work. These lantanas are extremely valuable in the xeriscape. They are heat, wind, drought, and quite salt tolerant. They thrive in poor soils or sand as long as good drainage and full sun are provided. They present a wealth of color, ranging from white to orange and red, with vibrant bicolors. All are valuable color accents in the xeriscape and may be used as bushy ground covers, borders, in planters, or containerized specimens for the sunny lanai. They will grow moderately rapidly to 2–4 feet in height. All parts of the plant are considered toxic.

Lantana montevidensis

ZONE 4 H(HPWRA)

TRAILING LANTANA P

Verbenaceae (Verbena Family)

A useful South American shrub, this lantana grows to 3 feet in height, spreading to 6 feet across. Its lavender flowers are produced almost all year. Other flower color forms include yellow, gold, white, and white and lavender. It does best in full sun in a well-drained soil. It is heat and drought tolerant and moderately salt tolerant. Use it as a ground or bank cover or bright accent where a splash of color is needed in the xeriscape. It can also be used in planter boxes and hanging baskets. There is evidence that lantanas contain a toxic chemical. Be careful not to chew or ingest any part of the plant or let animals eat any part of the plant. Because this species rarely sets seed, it is not considered invasive. All forms are moderate growers.

Lycium sandwicense

‘ŌHELO KAI

ZONE 4

Solanaceae (Potato Family)

Native to Hawai‘i as well as several eastern Pacific islands, this succulent, spreading shrub may reach 3 feet in height. It is wind, heat, salt, and drought tolerant. Its bright red fruit is prized by the lei maker. It will grow on rocky coastal ledges swept by salt spray, as well as in leeward coastal areas. It finds use in beach gardens as well as the xeriscape as a bushy ground cover and color accent. It is a somewhat slow grower.

Malvaviscus arboreus var. *drummondii*

SLEEPING HIBISCUS

ZONE 3

Malvaceae (Hibiscus Family)

Native to south central and coastal eastern Texas and into northern Mexico, this species shows moderate tolerance to drought and provides an attractive shrub reaching 5 feet in height. It is heat and wind tolerant and partially salt tolerant. Sleeping hibiscus is a moderate grower, thriving in full sun in any well-drained soil. It may be used as a colorful low hedge or border, as a bank cover, or as a tough potted specimen for a sunny deck. Flowers may be red, pink, or white and have been used in leis.

Myoporum sandwicense

CREEPING NAIO

ZONE 4

Myoporaceae (Myoporum Family)

Botanically the same as the native Hawaiian tree *naio* (it is listed under "Small Trees"), this decumbent form is a wide-spreading ground cover that may attain 2 to 3 feet in height. Its foliage is a medium green. Flowers are small. Creeping *naio* is wind, drought, and salt tolerant and makes an excellent ground cover or bank cover in the xeriscape and the beach garden. Give it full sun and a very well-drained soil. It will not tolerate soggy soil. It is a moderate grower.

Nototrichium divaricatum

NĀ PALI KULUI,
NĀ PALI ROCKWORT

ZONE 3

Amaranthaceae (Amaranth Family)

A rare, newly described *kului* from Kaua'i, this species grows to 5 or 6 feet in height, forming a dense shrub. Its leaves are a brilliant silvery white. It is heat, salt, and wind tolerant and shows moderate tolerance to drought. Use it as a bright accent among dark-foliaged plants or dark rocks or as an eye-catching, deep ground cover, low hedge, or screen. It produces a bright spot in the nightscape. Plant it in full sun in an open, well-drained soil. It is a fairly rapid grower.

Olea europaea 'Little Ollie'

DWARF OLIVE

ZONE 3

Oleaceae (Olive Family)

This is a wonderful dwarf form of the common olive. It carries silvery green to olive green foliage but slowly grows only to 6 feet in height with an equal spread. It does not flower. It tolerates poor soil, drought, heat, and wind but must have good drainage. Its dense foliage makes it an ideal subject for producing a hedge, low windbreak or screen, or a bright accent among dark green foliage or in the night garden with uplighting. It withstands clipping but works best when left in its natural form.

Pedilanthus macrocarpus

WAX PLANT **P**

ZONE 4

Euphorbiaceae (Euphorbia Family)

From Baja California and the west coast of Mexico, this husky species will reach 5 feet in height, with many green stems. It is highly drought, heat, and wind tolerant and moderately salt tolerant. It produces red flowers in late summer. Use it as a vertical accent above low ground covers or as a background hedge behind shorter, more colorful plants. It is used by flower arrangers. In its native areas, it is cooked in order to release a wax used for making candles. A euphorb, its sap is poisonous. Handle with care. It is a moderate grower.

Pelargonium x hortorum

GARDEN GERANIUM,
ZONAL GERANIUM

ZONE 2

Geraniaceae (Geranium Family)

Possibly one of the most popular bedding and potted plants throughout the world, the common geranium is the product of a complex of hybridizations from species native to South Africa. Easily grown in full sun and planted in a well-drained soil, it will reach 5 feet in height, forming a spreading shrub. Flowering is profuse, displaying an amazing array of colors from white to pink and salmon, orange, red, and violet, plus numerous bicolors. There are both double and single flowers. Some hybrids have leaves lightly banded with bronze. Deadheading greatly extends the flowering period, although in the subtropics flowering is year-round. Give these plants a well-drained soil and full sun. They are heat tolerant and moderately tolerant of wind and drought and are moderate growers.

Pittosporum tobira 'Wheeler's Dwarf'

WHEELER'S DWARF
PITTOSPORUM

ZONE 2 EVALUATE

Pittosporaceae (Pittosporum Family)

A dwarf form of the common Japanese pittosporum, this tough plant will form a dense mound of foliage to 3 feet in height with at least an equal spread. It grows best in full sun but will perform satisfactorily in light shade. It is not particular as to soil as long as it is well drained, but it does not do well in sand. It has good salt and wind tolerance and moderate drought tolerance. It makes a good ground or bank cover or border. It is a rather slow grower.

Aloe dawei

Plumbago zeylanica

WILD PLUMBAGO, ʻILIEʻE **P**

ZONE 4

Plumbaginaceae (Plumbago Family)

Native to Hawaiʻi and Old World tropics, this highly drought tolerant spreading shrub will slowly reach 3 feet in height. It is found in dry coastal areas, in soil and lava flows. It is heat and wind tolerant. It is an excellent ground or bank cover for a difficult place in the xeriscape. Its white flowers are small.

Plumeria 'Petite Pink'

ZONE 2

DWARF SINGAPORE, PETITE PINK PLUMERIA

Apocynaceae (Dogbane Family)

'Petite Pink'

'Petite Pink'

'Petite White'

'Petite Yellow'

Hybrid plumerias, probably complex *P. obtusa*/*P. rubra* crosses and selections, have produced at least three available dwarf forms with fragrant blossoms. Among these is 'Petite Pink', a small-flowered, attractive pink. Of two others without published names, one has white and the other yellow flowers. We will refer to them as 'Petite White' and 'Petite Yellow'. All three slowly form dense mounds of foliage up to 6 feet in height and spread. Flowering is quite profuse. All three are partially evergreen. Full sun, a well-drained soil, and space to spread are essentials. They are only moderately drought tolerant but both heat and wind tolerant. Use as a border, low hedge, in the rockery, or as a color accent in a small space.

Rosmarinus officinalis

ROSEMARY

ZONE 3

Lamiaceae (Mint Family)

'Prostratus'

Originating in sunny, dry areas of the Mediterranean, this excellent species has been cultivated since ancient times. Reaching 6 feet in height, all parts of the plant are used: leaves are used in cooking and flowers are processed to obtain an oil used in cosmetics and medicines. It is heat, wind, drought, and salt tolerant, finding use in the xeriscape as a flavoring, low hedge, or screen. Its plentiful blue flowers are a color accent. Sometimes listed in the trade as *R. officinalis* 'Prostratus', this is a popular creeping form of the common rosemary. It will reach 2 feet in height with an 8-foot spread. Use it in the xeriscape as a ground or bank cover, in the rockery, or cascading over the top of a low retaining wall. Its leaves may also be used in cooking. Both are moderate growers.

Russelia equisetiformis

CORAL PLANT,
FIRECRACKER PLANT

ZONE 3

Scrophulariaceae
(Snapdragon Family)

This Mexican species rather rapidly produces masses of light green, arching stems reaching 5 feet in height and length. They are brightened by an abundance of red tubular flowers, the whole giving the effect of a colorful waterfall. In full sun or light shade and given a well-drained soil, it will flower most of the year. It is moderately drought tolerant and wind and heat tolerant. There are forms with bright yellow and pink flowers. In the xeriscape use it as a color accent, bank cover, or wall top planting, or in a rockery. It is a moderate grower.

SANSEVIERIA SPECIES

There are about sixty species of *Sansevieria,* from tropical Africa, Madagascar, and Asia, ranging in height from a few inches to 8 feet. Foliage varies from dark to light green and to silvery green and may display variegations and stripes of yellow or white and may be broad, narrow, or cylindrical. They prefer full sun or light shade and a well-drained soil. They are heat, wind, drought, and salt air tolerant and will survive considerable gardeners' abuse. They are popular potted plants. Flowers are white, carried on stems rising a foot above the foliage, and are used in arrangements, as are the leaves. The leaves will form roots and may be replanted in the xeriscape. Sansevierias spread by surface rhizomes and become slowly invasive. A root barrier is useful in controlling their place in the garden. In the xeriscape they find use as a ground cover (see *S. trifasciata* Hahnii under "Ground Covers"), on a slope to control erosion, in the rockery as foliar/color accents, or in a mixed border. Not recommended are those species with cylindrical foliage, as they are tipped with a strong, sharp point and may pose a hazard to the gardener, his family, and visitors unless used far away from possible contact. They are slow growers.

Sansevieria dawei

DAWE'S BOWSTRING HEMP

ZONE 4

Agavaceae (Agave Family)

Reaching 3 feet in height, this tough species from Kenya and Uganda grows rather slowly, producing broad, attractive leaves. It is heat, wind, drought, and salt tolerant, making it a good choice for both the beach garden and the xeriscape. It provides a strong vertical accent, or it can be used in a border planting or on slopes for erosion control. It performs best in light shade or full sun planted in a well-drained soil.

Sansevieria trifasciata

BOWSTRING HEMP

ZONE 4 H(HPWRA)

Agavaceae (Agave Family)

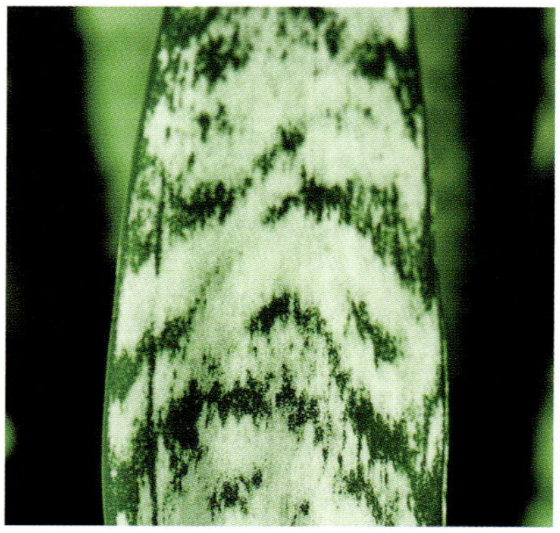

'Laurentii'

'Bantel's Sensation'

Sansevieria kirkii var. *pulchra*

Native to dry places in tropical Africa, this species will grow to 4 feet in height. A cultivar, *S. trifasciata* 'Laurentii', has yellow-margined leaves, while *S. trifasciata* 'Bantel's Sensation' produces leaves striped with light yellow. *S. trifasciata* 'Forescate' displays bright yellow leaves margined with dark green. All will grow in full sun or light shade and tolerate a variety of well-drained soils, heat, wind, salt air, and drought. All are nearly indestructible. Similar to bowstring hemp, *S. kirkii* subsp. *longifolia* grows to 8 feet in height, while *S. kirkii* var. *pulchra* produces leaves to 6 feet in height with wavy leaf margins and splashes of pale yellow. All are slow growers.

Santalum ellipticum

'ILIAHIALO'E, COAST SANDALWOOD

ZONE 4

Santalaceae (Sandalwood Family)

This unusual endemic shrub will slowly reach 2 feet in height and twice that in its spread. It is heat, wind, salt, and drought tolerant and the perfect candidate for the xeriscape and the beach garden. In the wild it is found growing from just above the reach of waves as well as inland in areas of strong, persistent onshore winds. It must have full sun and good drainage and should not be overwatered. Use it as a ground cover or accent among dark foliage. It is an excellent cover for rocky slopes. It is a slow grower.

Senecio cineraria

DUSTY MILLER

ZONE 3

Asteraceae (Sunflower Family)

A Mediterranean coast native, this species fares best at higher elevations, although it performs for short periods at low elevations if given a well-drained soil and full sun. It has good wind, heat, salt, and drought tolerance. It makes a fine border or bedding plant and its silvery white foliage can liven an otherwise dull planting. It is a moderate grower.

Senecio crassissimus

THICK LEAVED SENECIO

ZONE 4

Asteraceae (Sunflower Family)

This Senecio, native to Madagascar, will reach 2.5 feet in height and produces many vertical branches. Its flattened, fleshy leaves are a silver gray edged with purple. Stems are purplish. Flowers, held well above the stems, are bright yellow. It is heat, wind, and drought tolerant. Provide a well-drained soil in a very sunny spot. It may be used as a textural and color accent, ground or bank cover, or in a mixed border. It is a rather slow grower.

Serissa japonica

SERISSA

ZONE 3

Rubiaceae (Coffee Family)

'Variegata'

From Southeast Asia, this delicate-looking species will attain 3 feet in height. It is a slow grower, producing small white flowers from spring to fall. It shows good wind, salt, and drought tolerance. Use it as a fine-textured, untrimmed hedge or carefully shaped specimen in sun or light shade. It may be used as a bonsai subject. A well-drained soil is required. There is a green and white variegated leaf form, *S. japonica* 'Variegata'.

Sesbania tomentosa

'ŌHAI

ZONE 4

Fabaceae (Bean Family)

A rare and endangered Hawaiian species, this useful plant is found in hot, dry, coastal areas on all the main Hawaiian islands and Necker and Nihoa islands to the northwest. It grows best in direct salt wind. Usually seen as a low, widely spreading shrub reaching 3 feet in height, there are treelike forms reaching 18 feet in height. Foliage is silvery and silky, and flowers vary in color from salmon to orange red, scarlet, and yellow. The flowers are highly prized by the lei maker. Use it as a ground cover in beach gardens or as an accent against a lower, dark green ground cover. Plant it in full sun in perfectly drained soil or sand in a place where its flowers can be appreciated and picked. It is a moderate grower.

Sida fallax

'ILIMA

ZONE 3

Malvaceae (Hibiscus Family)

A highly variable native species, *'ilima* is also found on other Pacific islands and westward to China. A prostrate form, *'ilima kū kahakai,* which is both drought and salt tolerant, is discussed under "Ground

Covers." Away from strong onshore salt winds, it may grow to 5 feet or more in height, bearing flowers ranging from pale yellow to strong orange yellow and bronze red. They are about an inch in diameter. Five hundred flowers are needed to make a one-strand lei. All colors are prized by the lei maker. It is drought, heat, and wind tolerant. *'Ilima* frequently figures in Hawaiian myths and legends. It was planted around homes for its flowers and for medicinal purposes and is sometimes referred to as *'ilima 'apiki*. *'Apiki* means "mischievous," as it was believed that a lei *'ilima* attracted mischievous spirits. All forms are moderate growers.

Tagetes erecta

AFRICAN MARIGOLD

ZONE 2

Asteraceae (Sunflower Family)

Inasmuch as this species is native to Mexico and Central America, its common name is a puzzle. It was used ceremoniously by the Aztecs and up to this day in Mexico and Central America. It has many medicinal uses. Easily grown, it has become popular in India, where great garlands are made from its flowers for special occasions. A well-drained soil and full sun are needed for its best performance. It will grow to 3 feet in height. Horticulturists have done extensive work with the African marigold to produce flowers ranging from yellow to orange and white. Deadheading the flowers will extend the period of flowering. Commercially, in addition to its landscape color value, flowers are collected to extract lutein as a food color used to enhance the color of egg yolk and meat in poultry. In Europe it is used as a food colorant and for dyeing fabrics a golden yellow, orange, or olive green depending on the mordant used. Marigold roots are said to repel insects and nematodes. It is a fairly rapid grower. It is an annual.

Vitex rotundifolia

BEACH VITEX,
KOLOKOLO KAHAKAI

ZONE 4

Verbenaceae (Verbena Family)

Native to coastal areas of Hawai'i and throughout the Pacific and Indian oceans, this is an excellent choice for beach gardens with onshore winds. It shows excellent tolerance of salt, heat, wind, and drought. It grows to 2 feet in height at the beach, but under cultivation it may reach 4 feet in height. Provide a site in full sun and a well-drained soil, even pure sand. The foliage is pleasantly aromatic. Both the foliage and its bluish flowers are used by the lei maker. It is a moderate grower.

Wikstroemia uva-ursi

'ĀKIA

ZONE 3

Thymeleaceae ('Ākia Family)

An endemic Hawaiian species forming a dense shrub to 3 feet in height and with an equal spread, this tough plant grows well in a variety of well-drained soils. It prefers full sun but will perform satisfactorily in light shade. Its gray green foliage, small, spherical red fruit, and clusters of yellow flowers are used by the lei maker. It is wind, heat, salt, and drought tolerant and makes a fine bank or ground cover or used as a color accent for its fruit and for its foliage above low, dark green foliage. It is a moderate grower.

Zamioculcas zamiifolia

ARUM FERN

ZONE 3

Araceae (Aroid Family)

This species is native to dry grasslands and low forest areas of Zanzibar and Tanzania. It is a moderate grower, reaching 4 feet in height in rocky, full, or semishaded areas or in full sun with midday shade. It must have good drainage. It is an excellent interior plant, capable of withstanding abuse even if placed in fairly dark areas and with little water. It is a good xeriscape subject as a low hedge, as an accent in a partially shaded rockery, or as background for ground covers with small, light green foliage. It is heat, wind, and drought tolerant and moderately tolerant of salt.

Zinnia violacea

ZINNIA

ZONE 3

Asteraceae (Sunflower Family)

Although an annual rather than a shrub, we include this Mexican species as a valuable addition to the xeriscape palette. Rather rapidly reaching 3 feet in height, it bears an abundance of flowers in a wide range of colors from white and all the pinks, roses, and salmons to orange, yellow, red, lavender, and even green, as well as bicolors. There are also dwarf forms available, growing to only a few inches in height and displaying the same color range. There are both double and single flowers available. Provide a well-drained soil in full sun. Avoid wetting its leaves, as mildew may develop. Use soaker hoses at a low rate of flow or subsurface emitters. Zinnias prefer being kept on the dry side; shade and soggy soil must be avoided. In addition to their bright color accenting for the xeriscape, zinnias make excellent cut flowers. The dwarf forms are used in lei making. Deadheading will lengthen the flowering period. Zinnias are heat and wind tolerant and moderately tolerant of drought.

MEDIUM SHRUBS
6–10 FEET

Adenium arabicum

YEMEN DESERT ROSE P

ZONE 4

Apocynaceae (Dogbane Family)

Occurring along the southern and western margins of the Arabian Peninsula, this many-trunked, variable species reaches 10 feet in height depending on environmental conditions. Both leaves and flowers may vary from smooth to hairy and in size. They are borne abundantly and are usually a bright pink with a white center. It is a valuable addition to the xeriscape, as it shows a strong tolerance of drought, heat, wind, and salt. It produces an excellent splash of color and can also be used as a hedge or screen or tubbed specimen on a sunny deck. Its sap is toxic. It is a slow grower.

Adenium multiflorum

SABI STAR, IMPALA LILY P

ZONE 4

Apocynaceae (Dogbane Family)

Growing up to 10 feet in height, this floriferous succulent from the east coast of South Africa north to Mozambique and neighboring countries to the west provides masses of fragrant red and white flowers. Plants with pure white flowers are known. Petals are pointed. The central white portion of the flower is strongly star shaped. It requires very well-drained soils. Plant it in full hot sun. Like its other *Adenium* relatives, it is heat, wind, salt, and drought tolerant. It finds excellent use as a strong color accent in the xeriscape. Sabi star is deciduous, flowering for five months during cool weather when leafless. It produces an amazing splash of color and provides a seasonal change in the landscape. Its sap is toxic. It is a slow grower.

Adenium obesum

DESERT ROSE P

ZONE 4

Apocynaceae (Dogbane Family)

Growing to 10 feet in height (one reference indicates 12 feet), this succulent plant provides a wonderful display of color in the xeriscape. Its abundant flowers vary in color from red, reddish pink, and pink to white. It develops many trunks and branches from a broad, water-storing base. Flowers are borne abundantly on branch tips. The desert rose has become an extremely popular potted plant worldwide because it will tolerate considerable abuse, although it is not tolerant of cold. It is native to the broad area south of the Sahara from Senegal and Sudan to East Africa. It is drought, heat, wind, and salt air tolerant. A large number of named hybrids are on the market, which greatly adds to the available color range. It is probably the most colorful plant available for the very dry garden. Plant it in full sun in a very well-drained soil. Use it as a strong color accent in a low planting, as a tubbed accent for the terrace garden, or as a border or hedge. It is especially attractive in the rock garden. Use caution in handling: its sap is toxic but has medicinal uses. In its native areas, the sap is used to treat scorpion stings and snakebite. It is a rather slow grower. The Socotra desert rose, *A. obesum* subsp. *socotranum* (not figured), is a single-trunked tree up to 25 feet in height with multiple branching flowers at its tip. Flowers are light to medium pink. It is an extremely slow grower and recommended only for the hobbyist.

Adenium somalense

SOMALIA DESERT ROSE P

ZONE 4

Apocynaceae (Dogbane Family)

var. *nova*

A rather rapid grower, this Adenium produces a small tree reaching 10 feet in height (one reference indicates 15 feet in height). Native to very hot, dry areas of Somalia and south through the rift to Kenya and Tanzania, it provides another valuable plant for the xeriscape. Its flowers are smaller than other adeniums, rarely exceeding 2 inches across, and they vary in color from pink to deep red. Plant it in full sun in a good, well-drained soil. It is heat, drought, wind, and salt tolerant. Use this desert rose as a tall color accent or as a screen. Handle with care, as with other adeniums, its sap is toxic. In selecting plants, be certain that your purchase is from northwestern Kenya or Somalia to avoid planting a more shrubby variation native to its southern range. A new, highly attractive form recently discovered in Kenya grows to 15 feet in height with very narrow leaves and a spreading canopy. It is equally valuable in the xeriscape.

Agave americana

CENTURY PLANT, MAGUEY T

ZONE 4

Agavaceae (Agave Family)

Var. *marginata*

'Mediopicta Alba'

'Variegata'

Producing a large rosette of leaves up to 12 feet in diameter, this spectacular Mexican species has been carried around the tropical world where it is prized for its "explosive" form, its several variegations, and its seeming indestructibility. It is wind, heat, salt, and drought tolerant and grows in almost any soil (although faring best in an open, well-drained soil), but it must have full sun. Variegations include *A. americana* var. *marginata, A. americana* 'Mediopicta Alba', and *A. americana* 'Variegata'. All provide strong visual accents in the xeriscape. They work well on steep slopes, as barrier plants, and as color and textural accents. Due to its size, it is not normally recommended for the small garden. In any event, place it where its sharp leaf-tip spines will not come into contact with people. Another large agave, *A. sisalina,* is the source of sisal stalks used for fencing, for a strong fiber used to make string and ropes, and for carpeting and paper. It has been cultivated since ancient times. Several other species of maguey are the source of pulque, the fermented sap. If this is heated and distilled, it is *mezcal.* The sap of blue agave, *Agave webberi,* produces tequila. Known in Mesoamerica in prehistoric times, pulque became such an element socially, economically, and, as a consequence, religiously that myths, legends, and cults proliferated around it and its source, maguey. It was served to priests as a ritual intoxicant to increase their enthusiasm and to the victims of human sacrifice to ease their demise. All are moderate growers.

Agave attenuata

DRAGON TREE AGAVE, SWAN'S NECK AGAVE

ZONE 4

Agavaceae (Agave Family)

'Variegata'

A native of tropical Mexico, this agave develops a rosette of succulent, gray green, spineless leaves to 2.5 feet long carried on a trunk that may reach 5 feet in height. In summer a long, slender, curving stem up to 10 feet in height rises from the heart of the plant, bearing hundreds of greenish-white flowers. The base of the plants throws many small offsets that are used in arrangements. Give it full sun and a well-drained soil, but it is tolerant of poor soils as well as drought, salt, and wind. It is also fire resistant. In the xeriscape it offers a bright accent. A variegated form, *A. attenuata* 'Variegata', is available. It carries its bold leaves striped in yellow. There are many agave species available to the xeriscaper, almost all of them with sharp spines at leaf tips and along the sides of leaves. *Agave attenuata* is unique. It is a moderate grower.

Allamanda laevis

TREE ALLAMANDA

ZONE 3

Apocynaceae (Dogbane Family)

Slowly reaching 10 feet in height, this drought-tolerant species is native to hot, dry places in northeastern Brazil. Its few branches are capped with a handsome whorl of shiny green leaves and a terminal cluster of large, bright yellow flowers. It is best used as a background plant for color with shrubbier species in the foreground to hide its bare stems. It does best in a well-drained soil and planted in full sun. It is heat and wind tolerant. Avoid its white sap, which may cause eruptions on susceptible skins.

ALOE SPECIES

For common names, those marked with a single asterisk (*) are listed by the Institute for Aloe Studies; two asterisks (**) indicate names found online; and three asterisks (***) indicate that no common name was found by the authors.

Aloe arborescens

TORCH ALOE *

ZONE 4

Aloeaceae (Aloe Family)

'Variegata'

This species, which may reach 9 feet in height with many branches, is a popular xeriscape plant in South Africa. Its flowers are a strong red color. Its leaves produce a sticky sap that is under study for the treatment of x-ray burns. Readily hybridizing in the wild, fifteen natural hybrids are recognized. Also figured is a variegated form. It is heat, drought, and wind tolerant and shows moderate tolerance of salt. Plant it in a well-drained, open soil in full sun. Use it as a color accent, as a screen or hedge, or as an erosion control plant for a slope. It is valued in the rockery. Full sun and well-drained soil produce the best plants.

Aloe ferox

CAPE ALOE *

ZONE 4

Aloeaceae

With its scarlet or orange flowers on multiple spikes, this South African aloe provides striking color in the dry garden. It grows only a single trunk 7 to 10 feet in height topped by a large rosette of leaves up to 6 feet in diameter. Use it as a strong accent among smaller species. It shows off best backed by dark green foliage. In a multiple planting, it may form a screen or windbreak. It must have a well-drained soil and full sun. It is heat, drought, wind, and salt tolerant.

Aloe vaombe

MALAGASY TREE ALOE **

ZONE 4

Aloeaceae (Aloe Family)

Slowly growing to 10 feet in height, this single-trunked aloe produces a broadly branched inflorescence bearing a spectacular winter crop of bright red flowers. It is native to arid areas of south and southwest Madagascar and is heat, wind, and drought tolerant. Plant it in a well-drained soil in full sun. Use it as a striking color accent or as a hedge, screen, or windbreak.

Bauhinia grandidieri

DWARF MADAGASCAN BAUHINIA

ZONE 3

Fabaceae (Bean Family)

This rare species is native to xeric scrublands of south and southwest Madagascar. It reaches 10 feet in height, forming a somewhat delicate looking shrub with small leaves and many small, lilac pink flowers. It is a slow grower and likes full sun and a well-drained soil. Plant it close at hand where it can be appreciated. Use it in a mixed border, as an accent, or in the rockery. It is heat and wind tolerant and shows moderate tolerance of drought.

BIDENS SPECIES

There are nineteen endemic Hawaiian species of *Bidens*. Some inhabit dry areas and are subjects for the xeriscaper. These have been presented together to show how the Hawaiian bidens have developed into many useful landscape species.

Bidens cosmoides

PO'OLĀNUI

ZONE 2

Asteraceae (Sunflower Family)

This unusual species, named *po'olā nui,* is found on Kaua'i in mixed mesic forest. While performing best in cooler, moist situations, it flowers under rather dry, low-elevation conditions. It may reach 8 feet in height, abundantly producing its upside-down flowers, which are prized by the lei maker. It grows well in both full sun and light shade and prefers a rich, well-drained soil. Use it for its lei flowers, as a conversation piece, and as a colorful ground or bank cover. It is a moderate grower.

Bidens hillebrandiana

KO'OKO'OLAU

ZONE 2

Asteraceae (Sunflower Family)

This small species reaches 12 inches in height or may be decumbent. It is native to exposed coastal cliffs and bluffs of East Maui, Moloka'i, and Kohala. It shows only moderate tolerance of drought. It may be used as a ground cover, a color accent in the border, or a bank cover. Give it full sun in a well-drained soil.

Bidens mauiensis

KO'OKO'OLAU

ZONE 4

Asteraceae (Sunflower Family)

A low-growing species to just 1 foot in height, this bidens is found in dry coastal areas of Lana'i, Maui, and Kaho'olawe. Full sun and a well-drained soil are needed for good growth. Use it as a ground cover, as a border, or among rocks. Its clear yellow flowers will be a bright spot in the dry garden.

Bidens menziesii

KO'OKO'OLAU

ZONE 4

Asteraceae (Sunflower Family)

This larger species sometimes reaches 12 feet in height on arid leeward slopes of Moloka'i and West Maui and in dry areas along the Saddle Road between Mauna Kea and Mauna Loa on Hawai'i. A well-drained soil and full sun are needed for growth. It bears yellow flowers used in leis. It works well as a colorful screen or hedge or as background for lower shrubbery. Very similar is *Bidens menziesii* subsp. *filiformis,* native to similar areas of the Big Island. It is found on the lower, dry slopes of Hualālai almost to sea level. Both are heat, wind, and salt tolerant.

Bidens micrantha subsp. *ctenophylla* (not figured)

KO'OKO'OLAU

ZONE 4

Asteraceae (Sunflower Family)

This rare species found in very dry habitats at low elevations in Kealakehe on leeward Hawai'i may reach 9 feet in height, with a rather columnar growth. It produces masses of bright yellow flowers most of the year, followed by a large seed crop. Cuttings root readily and seed germination is excellent. This fine endemic should be brought into cultivation. It would make a good, colorful hedge, screen, or windbreak or a bright accent. It must have good drainage and full sun. It must not be overwatered. All are moderate growers.

Brighamia insignis

'ŌLULU

ZONE 3

Campanulaceae (Bellflower Family)

This is an endangered Hawaiian species native to sea cliffs just above the reach of waves to elevations of over a thousand feet on Kaua'i and Ni'ihau. Its large, succulent leaves are salt tolerant; the single succulent stem may reach 15 feet in height. Moderately drought tolerant, it requires perfect drainage and full sun. It bears whitish yellow, fragrant flowers throughout the year. They may be used to make a lei. Note: due to its shape, we are treating this species as a large shrub rather than a small tree. It is a slow grower.

Brighamia rockii

PUA 'ALA

ZONE 3

Campanulaceae (Bellflower Family)

Another endangered Hawaiian species very close to *B. insignis*, *pua 'ala* bears abundant pale yellow flowers described as "sweet scented with an odor like that of violets" (*pua 'ala* translates as "fragrant flower"). They may be used to make a lei. Native to sea cliffs of windward Moloka'i, *pua 'ala* is found just above the wave line to an elevation of 1,200 feet. Forming a single, unbranched succulent trunk to 15 feet in height bearing a head of large succulent leaves, it is highly salt tolerant and will tolerate moderate drought. It must have full sun and perfectly drained soil. In former times, Hawaiians in Wailau Valley cultivated *pua 'ala* around their homes. Note: due to its shape, we are treating *pua 'ala* as a large shrub rather than small tree. It is a slow grower.

Chadsia grevei

CHADSIA

ZONE 3

Fabaceae (Bean Family)

The extensive xerophytic scrublands of west and southwest Madagascar are home to this colorful shrub, which shows its abundant light tangerine orange flowers during warm months. It is a slow grower, reaching 10 feet in height. Found at low elevations and coastal dunes, it is heat, wind, drought, and partially salt tolerant. In its native areas, parts of the plant are used for medicinal purposes. It makes a fine screen or hedge as well as an eye-catching color accent.

Chrysobalanus icaco

COCOPLUM

ZONE 3

Chrysobalanaceae (Cocoplum Family)

Normally seen as a scrambling, dense beach shrub to 6 feet in height, the cocoplum, under ideal conditions, may reach as much as 30 feet in height. It is here treated for its value as a shrub in beach gardens and in dry situations. Plant it in well-drained soil or in sand in the most exposed part of the beach garden. It is wind, heat, drought, and salt tolerant. It bears edible fruit, and parts of the plant are used in its native range for medicinal purposes. It is native to coastal areas from south Florida to islands of the Caribbean, Mexico, and Brazil as well as to tropical West Africa. Use it as a low windbreak, hedge, or dune binder. One cultivar presents reddish colored new foliage and is a popular color accent. Cocoplum foliage is usually dark green, a visual relief in both the xeriscape and beach landscape where foliage may be grayish. It is a moderate grower.

Clerodendrum inerme

GLORY BOWER, SORCERER'S BUSH

ZONE 4

Verbenaceae (Verbena Family)

This remarkable shrub is native to the broad area from tropical Asia eastward to islands of the tropical Pacific. It associates with mangroves and will tolerate tidal flooding. Its abundant small, white flowers are fragrant. It is extremely heat, drought, wind, and salt tolerant. It grows to 10 feet in height and can be used as a dense bank cover or sand binder. It can be clipped into a hedge. Like many clerodendrums, however, it tends to sucker and is aggressive. Plant it where its spread can be contained. In its native areas it has many medicinal uses. It is a rather rapid grower.

Coprosma repens

MIRROR PLANT

ZONE 2

Rubiaceae (Coffee Family)

'Picturata'

'Variegata'

Native to New Zealand, this evergreen species does best in cooler elevations. Planted in full sun in a well-drained soil, it will form a dense shrub to 10 feet in height. It has moderate wind, drought, and salt tolerance. It makes a good screen or hedge. It is rather slow growing. There are several variegated cultivars available.

Cordeauxia edulis

JEHEB NUT N

ZONE 4

Fabaceae (Bean Family)

Native to the arid border region between Somalia and Ethiopia, the jeheb nut grows slowly to 8 feet in height, producing a dense, small-leaved shrub. Its bright yellow flowers are followed by a very nutritious nut with a chestnutlike flavor. Its dark green, glossy leaves are the source of a red dye as well as a tea. It is heat, wind, and drought tolerant. Give it a well-drained soil and full sun. Use it in the xeriscape as a low hedge or in a mixed border. It is excellent for the edible garden. Plant it where its flowers may be seen and its fruit picked. It is endangered in its native area.

Aloe ferox

Crassula arborescens

SILVER JADE PLANT

ZONE 4

Crassulaceae (Stonecrop Family)

'Tricolor'

This South African succulent will rather slowly reach 10 feet in height, forming a dense shrub that is tolerant of wind, heat, salt air, and drought. It performs best in full sun and a well-drained soil. It does not tolerate soggy soil. Use it as a hedge, windbreak, or as an accent in a planting of dark green, fine-foliaged species. An attractive variant, *C. arborescens* 'Tricolor', may be used as a subtle color accent. Silver jade is sometimes confused with its close relative, *Crassula ovata,* but it has larger, silvery, slightly cupped leaves. It is a good choice for both the dry garden and the beach garden.

Euphorbia coerulescens

BLUE EUPHORBIA **P S**

ZONE 4

Euphorbiaceae (Euphorbia Family)

Growing to 6 feet in height, this South African euphorb develops a multibranched clump. It is an attractive blue green. This plant is drought, heat, and wind tolerant and finds its use in the xeriscape as an accent plant, where its shape and color may be focalized. It will grow in any well-drained soil in full sun. Plant it where it cannot spread, as it develops rhizomes that may grow into new plants. With uplighting, it can provide dramatic color in the night garden. It is a rather slow grower.

Ficus deltoidea

MISTLETOE FIG

ZONE 2

Moraceae (Mulberry Family)

Growing slowly to a height of 10 feet, this Malaysian evergreen shrub may be used as a hedge, screen, or as a potted specimen for the lanai. It prefers light shade and is moderately drought and salt tolerant.

Furcraea foetida 'Mediopicta'

STRIPED FURCRAEA **S**

ZONE 4

Agavaceae (Agave Family)

A rather spectacular plant, this species from dry areas of Central America and northern South America produces leaves up to 7 feet in length, providing a spreading rosette of foliage 14 feet across. Leaves are spine tipped. Its inflorescence carries abundant flowers of creamy white on a stem that may reach 40 feet in height. It dies after flowering, usually 12–15 years after planting. Numerous offsets are produced on the inflorescence. It is heat, wind, and drought tolerant. The leaves of the species are green and are harvested for their useful fiber. As with other agaves, this species is probably tolerant of moderate salt wind and should be tried in beach gardens. It grows best in a good, well-drained soil but will tolerate sandy soils. It must have plenty of space and full sun. It is a textural and color accent and a bright focal point when uplighted in the night garden. All agaves are moderate growers.

Galphimia gracilis

GALPHIMIA, SPRAY OF GOLD, RAIN OF GOLD

ZONE 2

Malpighiaceae (Malpighia Family)

Reaching 9 feet in height, this evergreen species from Mexico and Central America produces its fragrant yellow flowers throughout the year. It performs best when planted in full sun in a well-drained soil. It has moderate drought tolerance but poor salt tolerance. Use it as a colorful hedge or screen or as a background for lower-contrasting colorful species. It is a moderate grower.

Hibiscus brackenridgei subsp. *brackenridgei*

MA'O HAU HELE

ZONE 3

Malvaceae (Hibiscus Family)

This attractive, rare, endangered shrub reaches 10 feet in height, but under optimum conditions it may be found as a small tree to 15 feet in height. It is found on all the main islands except Ni'ihau and Kaho'olawe, growing in lowland dry forest and shrubland. It bears large yellow flowers that may or may not carry a maroon spot at the base of each petal. The flowers open from noon to late afternoon. It is a fairly rapid grower given full sun and a very well-drained soil. *Ma'o hau hele* is the official state flower of Hawai'i. It is heat and wind tolerant and moderately drought tolerant and can be used in the xeriscape as a color accent, hedge, or screen. Closely related is *H. brackenridgei* subsp. *mokuleianus,* native to dry habitats on Kaua'i and O'ahu. It grows into a small tree. They are moderate growers.

Ilex vomitoria 'Nana'

DWARF YAUPON HOLLY

ZONE 2

Aquifoliaceae (Holly Family)

As indicated, this is a dwarf form of the Yaupon Holly. See "Small Trees" for general characteristics and horticultural requirements. Usually seen as an 18-inch-high shrub with a 3-foot spread, old specimens have been known to reach 6 feet in height with equal spread. It is a slow grower. Plant it in a well-drained soil. It is useful as a tall ground or bank cover, as a barrier hedge, screen, or low windbreak. A similar dwarf cultivar is 'Stokes Dwarf', with smaller, darker green foliage and more upright growth than 'Nana'. It is also known as 'Schilling's Dwarf'.

Juniperus chinensis 'Pfitzeriana'

PFITZER JUNIPER

ZONE 3

Cupressaceae (Cypress Family)

Rather rapidly growing to 10 feet in height with a 20 foot spread, this conifer, probably originating in China, shows good salt and wind tolerance and moderate drought tolerance. It makes an excellent deep bank cover, hedge, or windbreak where space permits or as a specimen or accent above smaller plants.

Kalanchoe beharensis

VELVET LEAF, FELT BUSH

ZONE 4 EVALUATE

Crassulaceae (Stonecrop Familiy)

'Fang'

Reaching 10 feet in height with at least an equal spread, this tough, succulent Madagascan shrub is a valuable candidate for the xeriscape. It has good heat, drought, wind, and salt tolerance. Use it as a striking accent where its rich color and velvety texture may be enjoyed, or as a hedge, windbreak, or in the rockery. Plant it in full sun in a well-drained soil. It is a rather slow grower. It is good for the night garden.

Leucophyllum frutescens

TEXAS RANGER

ZONE 3

Scrophulariaceae (Snapdragon Family)

'Silver Cloud'

'White Cloud'

Native to arid areas of Texas and Mexico, this tough species may reach 10 feet in height. It is slow growing, forming a compact shrub in full sun in a well-drained soil. It flowers abundantly in summer. There are several named forms with different flower colors, ranging from violet to purple to pink and more silvery foliage. It is heat, wind, and drought tolerant. Use it as a hedge or screen either clipped or pruned to natural shapes. Its flowers and foliage qualify it as a color accent. It is good for the night garden and forms a spectacular, bright accent when lighted.

Nerium oleander 'Petite Pink'

DWARF OLEANDER **P**

ZONE 4

Apocynaceae (Dogbane Family)

There are several dwarf oleanders growing to 10 feet in height, bearing small flowers and foliage. 'Dwarf Peach' displays peach-colored flowers, 'Little Red' shows red flowers, while 'Apricot Form' and 'Petite Salmon' bear, respectively, apricot- and salmon-colored blooms. They are heat and wind tolerant and moderately drought tolerant and are well used in the xeriscape as hedges, windbreaks, massed color accents, or individual color accents. They will tolerate salt air but not direct exposure to onshore winds. Flowers are borne on new growth. Spent stems should be cut back to the ground, although one reference recommends pulling off old canes rather than cutting. Its white sap may cause skin irritation. Immediately wash it off if contacted. Provide full sun and a well-drained soil for optimal results. They are moderate growers.

Nototrichium sandwicense

KULUʻĪ

ZONE 4

Amaranthaceae (Amaranth Family)

An endemic species, this Hawaiian shrub grows to 10 feet in height, displaying bright silvery green foliage. Plant it in a well-drained soil in full sun. It has good wind, drought, and salt tolerance. Use it as a light-colored hedge backing shorter dark green foliage, as a windbreak, or as an accent. It is especially attractive in the night garden. It is a moderate grower.

Osteomeles anthyllidifolia

ʻŪLEI

ZONE 3

Rosaceae (Rose Family)

Native to all the main Hawaiian Islands except Niʻihau and Kahoʻolawe, this tough shrub may reach 10 feet in height, with an erect, woody trunk. It is usually seen as a rambling shrub of smaller stature. It is also found in Tonga and the Cook Islands. It has many traditional uses: its white flowers and fruit are used by the lei maker; its hard wood was employed as digging sticks, fishing spears, and hoops for fishnets. It is heat, wind, salt, and drought tolerant. Use it as a bank or ground cover. It is attractive when used among dark lava rocks. It will cascade over a retaining wall. It is a moderate grower.

Pedilanthus tithymaloides subsp. *tithymaloides*

RED BIRD CACTUS, SLIPPER PLANT **S**

ZONE 4

Euphorbiaceae (Euphorbia Family)

Pedilanthus tithymaloides subsp. *tihymaloides*

Pedilanthus tithymaloides subsp. *tihymaloides*

Pedilanthus tithymaloides subsp. *padifolia*

Pedilanthus tithymaloides subsp. *padifolia*

Pedilanthus tithymaloides subsp. *smallii*

Pedilanthus tithymaloides subsp. *smallii*

This subspecies is an all-green plant from dry areas of Mexico to Suriname. It is a highly drought tolerant shrub growing to 6 feet or more in height. It is also salt and wind tolerant. Contrary to its common name, it is not a cactus. Although preferring a well-drained soil in a hot, sunny spot, it will tolerate a wide range of conditions. Frequently seen as a potted specimen in difficult situations, *Pedilanthhus* works well in the xeriscape as a hedge or accent plant in an area with low ground cover plants. More commonly seen in gardens is *Pedilanthus tithymaloides* subsp. *padifolia* with rounded leaves. It is native to islands in the Caribbean and may be used as a color accent. A variegated form is also available. 'Devil's Backbone' or 'Jacob's Ladder' are names given to *Pedilanthus tithymaloides* subsp. *smallii,* native to the Florida Keys and northern Cuba. It has markedly zig-zag stems. A cultivar, 'Variegatus', shows stems and leaves with white and red coloration. A dwarf form, 'Nana Compacta', is a dark green, many-branched plant reaching only a foot in height. It is not figured. Use care in handling all *Pedilanthus.* Members of the euphorbia family, they have caustic sap that may irritate sensitive skin. Use alcohol to remove sap from skin. It is not water soluble. Keep it away from eyes. They are all rather slow growers.

Petrea sp.

PETREA SHRUB

ZONE 3

Verbenaceae (Verbena Family)

Native to dry areas of Brazil, this relative of the sandpaper vine produces one or more slender, vertical stems reaching 6 feet in height. Each is capped with a colorful, almost horizontal spike of striking blue flowers. There is also a white flowering variation. It grows best in full sun in an open soil but will perform in light shade. It is heat and wind tolerant and moderately drought tolerant. It is a slow grower. Use it in the xeriscape as an unusual color accent or in a mixed border backing lower shrubs.

Pittosporum tobira 'Variegata'

VARIEGATED JAPANESE PITTOSPORUM

ZONE 2 EVALUATE

Pittosporaceae (Pittosporum Family)

Growing to 10 feet in height, this plant produces a dense shrub that has moderate tolerance of drought and salt and good tolerance of wind. It prefers a well-drained soil in full sun. It can be used as an accent specimen, as a hedge or windbreak, and can be used in arrangements. It is a moderate grower.

Plumbago auriculata

BLUE PLUMBAGO, CAPE PLUMBAGO, CAPE LEADWORT

ZONE 3

Plumbaginaceae (Plumbago Family)

'Alba'

'Midnight Blue'

Forming a dense, spreading mound 6 feet in height, this species provides excellent color in the xeriscape. It flowers heavily most of the year. Flower color is a sky blue. There is a white-flowered form as well as a dark blue form, 'Midnight Blue'. They are used to make leis. Plant it in full sun in a well-drained soil. It has moderate salt and wind tolerance and good heat and drought tolerance. Use it as a colorful bank cover, a color accent, or as a low hedge. They are rapid growers.

Polyscias fruticosa

PARSLEY PANAX, MING ARALIA

ZONE 2

Araliaceae (Panax Family)

This popular evergreen shrub native to the area from India to southern Polynesia will attain 10 feet in height. Its branching is vertical. Its foliage is dense. It is used as a specimen accent against a wall, as a hedge or screen, or as an excellent tubbed specimen. It is wind and heat tolerant but only moderately tolerant of drought and salt wind. It fares best in a rich, well-drained soil in part shade. It is a rather slow grower. Cut stems are used in arrangements, the foliage in leis.

Pseuderanthemum carruthersii

FALSE ERANTHEMUM

ZONE 2

Acanthaceae (Acanth Family)

var. *atropurpureum*

var. *atropurpureum*

var. *reticulatum*

var. *variegatum*

var. *variegatum*

Three varieties of this shrub are popular in Hawaiian gardens. Purple false eranthemum (*P. carruthersii* var. *atropurpureum*) grows to 10 feet in height, producing foliage irregularly variegated in dark purple, gray green and pink. Flowers are rose-purple. Yellow-veined false eranthemum (*P. carruthersii* var. *reticulatum*) grows to 6 feet in height, with bright yellow new foliage turning medium green upon maturing. Leaves are netted with yellow veins. Flowers are white dotted with purple. Variegated false eranthemum (*P. carruthersii* var. *variegatum*) grows to 8 feet in height, bearing dark green foliage irregularly variegated with gray green and yellow. Flowers are white dotted with purple. All grow well in full sun or light shade and, although preferring a moist soil, show moderate tolerance to drought and salt. They provide high color and may be used as low hedges or borders or as strong color statements. They are native to the southwestern Pacific and are moderate growers.

Punica granatum 'Nana'

DWARF POMEGRANATE

ZONE 3

Punicaceae (Pomegranate Family)

This showy, ever-blooming dwarf form of the common pomegranate will grow to 6 feet in height. It prefers full sun in a well-drained soil and is heat, wind, and moderately drought tolerant. It makes an excellent color accent or can be used as a border or low hedge. It is a slow grower.

Punica protopunica

SOCOTRA POMEGRANATE

ZONE 4

Punicaceae (Pomegranate Family)

From the island of Socotra, this is a primitive form of the common pomegranate. It grows into a shrub 10 feet in height. Its flowers are pinkish; its fruit yellow. It is heat, wind, and drought tolerant and provides an attractive accent in the xeriscape. It may be used as a hedge or screen, or it makes a good background for large-foliaged species. It is a moderate grower.

Rhaphiolepis indica

INDIAN HAWTHORN

ZONE 3

Rosaceae (Rose Family)

Native to southern China, this spreading evergreen shrub will reach 7 feet in height. Use it as a tall bank or ground cover or informal hedge. Plant it in full sun or light shade in a well-drained soil. It has moderate drought and salt tolerance and good wind tolerance. Flowers, fruit, and foliage are used in arrangements. It is a moderate grower.

Rhaphiolepis umbellata var. *integerrima*

YEDDO HAWTHORN, KOKUTAN

ZONE 3

Rosaceae (Rose Family)

This spreading, evergreen Japanese shrub will grow to 10 feet in height. Its white flowers are produced in winter and are followed by clusters of black fruit. Flowers, fruit, and foliage are prized by the arranger. Plant it in a fertile, well-drained soil in full sun. It has good wind, salt, and drought tolerance. Use it as a specimen accent or in a mass planting to provide deep ground or bank cover. It is a moderate grower.

Rondeletia odorata

RONDELETIA

ZONE 3

Rubiaceae (Coffee Family)

In spite of its Latin name, this shrub from Cuba and Panama does not bear fragrant flowers but earns its landscape kudos with masses of orange and red blossoms. Usually seen as a shrub 6 feet in height, it may, under ideal conditions, reach 10 feet in height. It thrives in full sun in a well-drained soil. It is wind tolerant and moderately drought tolerant. It makes a wonderful color accent either as a specimen or used in a hedge or low screen. It is a moderate grower.

Sanchezia speciosa

SANCHEZIA

ZONE 2

Acanthaceae (Acanth Family)

Thriving in both sun and light shade in a well-drained soil, sanchezia is an evergreen shrub from northern South America. It will grow to 8 feet in height and finds use as a screen, hedge, or a striking color accent. It displays moderate salt, wind, and drought tolerance. It is a rather slow grower.

Brighamia rockii

Scaevola taccada

BEACH NAUPAKA, NAUPAKA, NAUPAKA KAHAKAI

ZONE 4

Goodeniaceae (Naupaka Family)

Growing to 10 feet in height with an equal spread, this species is native to Hawaiian coastal areas as well as throughout the tropical Pacific. It thrives in pure sand or soil, even in hard mudflow rock in areas with strong onshore winds, as well as in leeward sites. It is strongly wind, heat, salt, and drought tolerant. It is an excellent sand binder and makes a good hedge or screen. Branches and large leaf rosettes are used in arrangements, while its white flowers and fruits as well as small leaf rosettes are used by the lei maker. It is a rather rapid grower.

Strelitzia reginae

BIRD OF PARADISE, CRANE FLOWER

ZONE 3

Strelitziaceae (Bird of Paradise Family)

var. *juncea*

var. *juncea*

A slow growing, clump-forming, evergreen South African plant reaching up to 6 feet in height, this species produces its flamboyant flowers all year. Select only those plants that bear flowers well above the foliage. It flowers best in full sun in a good, well-drained soil. It has moderate salt, wind, and drought tolerance. Both flowers and foliage are used in arrangements. It may be used as a bright color accent alone or in scattered clumps or as a hedge or screen. Not figured is a form with yellow flowers. A variant, *Strelitzia reginae* var. *juncea,* produces the same flamboyant flowers but has tubular, "rush-shaped" leaves terminating in a tough, sharp point. It is a strong color accent and its foliage provides a striking textural accent. It is recommended that gardeners take care in placing this form to prevent accidental contact with children and visitors, as well as the gardener.

Synadenium grantii

AFRICAN MILK BUSH s

ZONE 4

Euporbiaceae (Euphorbia Family)

'Rubrum'

'Rubrum'

Growing to 10 feet in height, this tropical East African, succulent shrub is very drought, heat, salt, and wind tolerant. Foliage is either green or reddish green. Care should be taken when handling this species, for its milky sap is quite irritating, although having medicinal properties. As soon as possible, wash off sap with alcohol. It is not water soluble. Plant it in full sun in a soil with good drainage. In the xeriscape it may be used as a hedge, screen, or windbreak. The form with reddish foliage is a good color accent.

Tecoma capensis

CAPE HONEYSUCKLE, I'IWI HAOLE

ZONE 3

Bignoniaceae (Catalpa Family)

'Aurea'

Mounding to 10 feet in height or more, this is a rambling shrub or vine native to South Africa. It flowers most of the year, producing bright orange blooms. There are also forms with yellow, gold, or apricot flowers. It must have good drainage and full sun. It is tolerant of heat, wind, and salt air and is moderately drought tolerant. It makes a good hedge if pruned regularly, an excellent bank cover, and a brilliant color accent used either as a specimen or mass planting. It may be used in a protected area in beach gardens. It is a rapid grower.

LARGE SHRUBS OVER 10 FEET

Acalypha wilkesiana

COPPER LEAF, JACOB'S COAT, BEEFSTEAK PLANT P

ZONE 2

Euphorbiaceae (Euphorbia Family)

'Picotee'

This large, sprawling evergreen shrub from Southeast Asia bears red or green and yellow variegated foliage. An interesting cultivar, 'Picotee Acalypha' (*A. wilkesiana* 'Picotee'), has green and white variegated foliage with a white-fringed leaf. Growing to 15 feet in height, they develop their color best in full sun but will thrive in light shade. They show moderate salt and drought tolerance. Use them as a very colorful hedge or screen or as a color accent among other less colorful plants. They are moderate growers.

ALOE SPECIES

For common names, those marked with a single asterisk (*) are listed by the Institute for Aloe Studies; two asterisks (**) indicate names found online; and three asterisks (***) indicate that no common name was found by the authors.

Aloe africana

AFRICAN ALOE **

ZONE 4

Aloeaceae (Aloe Family)

Usually producing a single-trunked plant, this South African aloe will slowly reach 12 feet in height. It is heat, wind, and drought tolerant. Long spikes of colorful yellow to yellow orange flowers appear in winter, making a bright accent in the xeriscape. Use it as a colorful accent or in groups to form an irregular hedge, screen, or windbreak. Give it a well-drained soil in full sun. It is a rather slow grower.

Aloe castanea

CHESTNUT ALOE **

ZONE 4

Aloeaceae (Aloe Family)

Named for its unique flowers, which are chestnut (castanea) in color, this South African native reaches 15 feet in height and may be a single or branched trunk bearing as many as twenty crowns. It may also branch basally. It is a striking, unique color accent for the xeriscape and may also be used as a hedge, screen, or windbreak. It requires full sun and a well-drained soil. It is a rather slow grower. It is heat, wind, and drought tolerant.

Aloe excelsa

ZIMBABWE ALOE ••

ZONE 4

Aloaceae (Aloe Family)

This species' natural range is quite widespread, from Zambia, Zimbabwe, and Malawi east to Mozambique. It slowly produces a 15-foot single trunk capped by a many-branched inflorescence bearing bright red and red orange to orange flowers. It is wind, heat, and drought tolerant and performs best in a well-drained soil in full sun. It finds use as a bright color accent or may be grouped to form a hedge, screen, or windbreak. It is a slow grower.

Aloe kedongensis

KENYAN ALOE ••

ZONE 4

Aloeaceae (Aloe Family)

Native to dry parts of Kenya, this aloe will reach up to 12 feet in height, forming a many-stemmed, dense shrub. It displays abundant, attractive scarlet flowers in winter. It requires full sun and does best in an open, well-drained soil. It is heat, wind, and drought tolerant. In the xeriscape, use it as a tall, singular accent or as a screen, hedge, or windbreak. It is a slow grower.

Aloe marlothii

MOUNTAIN ALOE •

ZONE 4

Aloeaceae (Aloe Family)

One of the tallest of the aloes, this single-trunked South African will reach 15 to 24 feet in height, topped by a rosette of leaves up to 9 feet across. Its abundant bright flowers are orange to yellowish orange in color, making a striking statement in the xeriscape. It fares best in full, hot sun in a well-drained soil. Use it as a bold accent singly, in a grouping, or in a multiple planting as a screen or windbreak. It is heat, wind, drought, and salt tolerant. It has a wide distribution: it is also found in Natal, Botswana, and Swaziland.

Aloe volkensii

•••

ZONE 4

Aloeaceae (Aloe Family)

Usually producing a single trunk to 15 feet in height or more, specimens developing one or two basal trunks are known. The trunk is capped by a large rosette of leaves 4 feet or more across and a many-branched inflorescence showing abundant flowers ranging from pale scarlet to bright reddish orange. The inflorescence grows to 2.5 feet above the foliage. It is heat, wind, and drought tolerant and may be used in the dry garden as a strong vertical accent or in informal groupings to form a hedge, screen, or windbreak. It requires full sun and a well-drained soil and is a moderate grower. This aloe is native to Tanzania, Kenya, and Uganda.

Bidens menziesii

see previous chapter

Bixa orellana

LIPSTICK PLANT, ACHIOTE, 'ALAEA

ZONE 2

Bixaceae (Annatto Family)

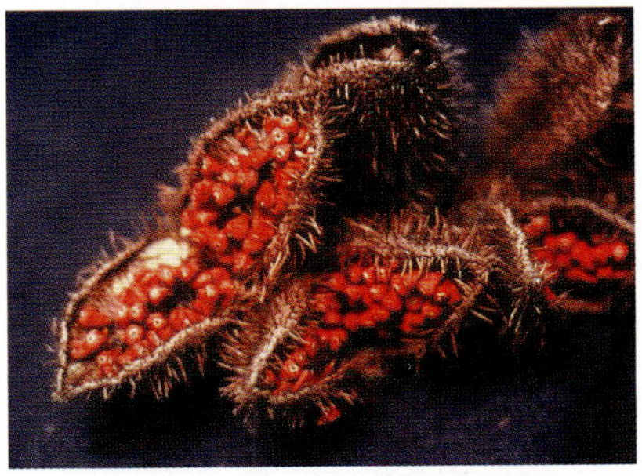

Growing to 30 feet in height, this tropical American shrub is useful as a hedge or screen and may be easily pruned into a small tree. It is prized for its colorful pods, which may be bright red, dull red, or yellow and are used in wreaths and table decorations. A bright red powder within the pod was used by Indians for face painting and is now employed as a coloring agent for oleomargarine. It is not particular about soil as long as it is well drained. It needs full sun for best pod coloring. Bixa is a moderate grower. Use it as an accent or screen. It shows low salt and wind tolerance but is moderately drought tolerant.

Bontia daphnoides

BONTIA

ZONE 4

Myoporaceae (Myoporum Family)

A remarkably tough, dense shrub, Bontia is wind, heat, drought, and salt tolerant. It will reach 25 feet in height. Plant it in full sun in any well-drained soil. A fairly rapid grower, pruning will assist in producing a tree form if so desired. Or use it unpruned as a windbreak, screen, or tall hedge. Bontia is native to hot, dry, coastal places from the Bahamas, Cuba, and other islands of the Caribbean to Guyana. Use it in the xeriscape and the beach garden.

Breynia disticha 'Roseopicta'

SNOWBUSH, SWEETPEA BUSH, CALICO PLANT

ZONE 2

Euphorbiaceae (Euphorbia Family)

Growing to 10 feet in height, this evergreen Melanesian shrub forms a dense hedge or screen. Its foliar colors of green, white, and red develop best in full sun but are satisfactory in light shade. It is only moderately drought tolerant but not salt tolerant. Plant it where its tendency to spread by suckering can be controlled. It is a moderate grower and an excellent color accent.

Brunfelsia americana

LADY OF THE NIGHT P

ZONE 3

Solanaceae (Potato Family)

Native to dry areas of the West Indies, this large shrub will reach 16 feet in height. It is bushy in habit but may be trained to a tree shape. It bears large white flowers that are strongly fragrant at night. It is moderately drought tolerant. Plant it where its fragrance may be enjoyed in the evening. It also finds use as a hedge or screen. Give it an open, well-drained soil. It is a moderate grower.

Calliandra haematocephala

RED POWDERPUFF, LEHUA HAOLE N

ZONE 2

Fabaceae (Bean Family)

Native to Bolivia, this rapid-growing evergreen shrub will grow to 18 feet in height with a 20-foot spread and is readily pruned into a small tree. It bears showy red or reddish pink flowers in fall and winter. There is a white-flowered form. It prefers full sun in a good, well-drained soil. It makes a good screen or enclosure or a colorful accent. Although usually associated with moisture, it shows adaptability by tolerating moderate drought once well established.

Calotropis gigantea

CROWN FLOWER, GIANT MILKWEED, PUA KALAUNU P+ S+

ZONE 4

Aesclepidaceae (Milkweed Family)

C. procera

Native to dry places from India to Southeast Asia, this species grows to 15 feet in height, preferring full, hot sun. It will grow in any well-drained soil, even sand, and is highly heat, drought, salt, and wind tolerant. Use it as a windbreak or screen or pruned into a small tree. Plant it where its white or lavender flowers may be easily picked; it is very popular with lei makers. Take care to prevent its milky sap from touching bare skin. It may cause a bad skin reaction. In India the flower is sacred to Shiva, and its flower buds form one of the darts of Kama, the God of Love, an Indian Dan Cupid, if you will, who shoots his arrows into the hearts of mortals. A slightly smaller species, *C. procera*, the small crown flower, H(HPWRA), has purple and white flowers. Uses, growth needs, and tolerances are the same as its larger relative. Both are moderate growers.

Carissa macrocarpa

NATAL PLUM T

ZONE 3

Apocynaceae (Dogbane Family)

An evergreen, South African species growing rapidly to 20 feet in height, the Natal plum bears bright red plumlike fruits used to make a delicious jelly. Its milky sap is nontoxic. Although tolerating light shade, it performs best in full, hot sun in any well-drained soil, even sand. Its white, fragrant flowers appear most of the year. It makes an excellent barrier hedge, tall screen, or windbreak and can be pruned into a small tree. Foliage, flowers, and fruit may be used in arrangements. Natal plum is highly heat, wind, salt, and drought tolerant. It is excellent for both dry and beachside gardens.

Codiaeum variegatum

CROTON

ZONE 3

Euphorbiaceae (Euphorbia Family)

'Gold Dust'

Reaching 20 feet in height, this popular Melanesian shrub is among the most colorful and variable plants available. Foliage may vary in color from yellow, green and yellow, and orange to pink and many shades of red and purple red, and it may be large-leaved, small-leaved, narrow to broad-leaved, and even curly. It makes an excellent hedge, screen, or windbreak. It is suggested that only one color and leaf shape be used for these purposes to avoid a rather garish Joseph's coat appearance. Specimen plants make startling color accents. Croton is happy in both full sun or light shade and, once well established, shows moderate drought and salt tolerance and good tolerance to wind. It can be pruned into a tree shape, providing a strong color focal point in the xeriscape. They are slow growers.

Didierea madagascariensis

OCTOPUS TREE

ZONE 4

Didiereaceae (Didieria Family)

Native to the unique "spiny desert" of southwest Madagascar, where 95 percent of the plants of that ecosystem are endemic, this species will slowly attain 15 feet in height. It is easily grown and develops several erect, slender trunks, giving rise to its common name. It is heat, wind, and drought tolerant. Give it a place in the xeriscape in full sun and a well-drained soil. Do not overwater.

Use it as a vertical accent, as a light screen, as an accent among other species with larger, green foliage, or as a silhouette accent against a wall. Its structure is dramatized by uplighting in the night garden.

Dodonaea viscosa

'A'ALI'I

ZONE 4

Sapindaceae (Soapberry Family)

Usually seen as a large shrub, this native Hawaiian species can reach 20 feet in height. A slow grower, judicious pruning will produce a treelike shape. It is drought, heat, wind, and salt air tolerant and finds good use in the xeriscape as a screen, windbreak, small tree, or background plant for showier species. It grows in almost any well-drained soil and does best in full sun. Its papery fruits vary in color from white to red to mahogany and are popular for lei making.

Duranta erecta

GOLDEN DEWDROP, DURANTA, PIGEON BERRY P

ZONE 3 H(HPWRA)

Verbenaceae (Verbena Family)

var. *alba*

'Variegata'

'Goldedge'

'Gold'

'Geisha Girl'

Reaching 20 feet in height, this species from the southern United States south to Argentina flowers abundantly, producing blue violet blooms followed by a heavy crop of attractive pendant clusters of golden fruit. There is a white-flowered variety, *Duranta erecta* var. *alba*. There are also several attractive cultivars: 'Variegata' with green and white variegated foliage; 'Goldedge' carries green leaves bordered with yellow and with purple flowers; 'Gold' produces golden yellow leaves and lavender blue flowers; and 'Geisha Girl' has dark green foliage and blue and purple flowers. They all thrive in well-drained soil in full sun. Use golden dewdrop as a hedge, screen, or windbreak or trained into a small tree shape. 'Gold' is a strong color accent. They show heat and wind tolerance and moderate drought tolerance, but no salt tolerance. All parts of the plant are toxic. They are fairly rapid growers.

Euphorbia lactea

MILK-STRIPED EUPHORBIA T P S

ZONE 4

Euphorbiaceae (Euphorbia Family)

'Cristata'

var. *variegata* 'White Ghost'

A tough euphorb, possibly from India but found now throughout the dry tropics, this species grows to 20 feet in height. A lighter green stripe on its darker green branches has made it popular with local gardeners. It is drought, heat, salt, and wind tolerant and excellent for the xeriscape. Its overall shape is somewhat columnar, producing limited shade. A variegated form, *E. lactea* var. *variegata* 'White Ghost', is an all-white, almost silvery form, making a striking, bright accent in the xeriscape and especially attractive with lighting in the night garden. New branches sometimes revert to green and should be removed. This euphorb forms an excellent screen or windbreak. Because it bears sharp spines, it should be planted where the gardener and his children and visitors cannot brush against it. Its sap may be irritating. Wash it off with alcohol as soon as possible; it is not water soluble. Full, hot sun and a well-drained soil produce optimum growth. A crested form, *E. lactea* 'Cristata', is prized by collectors and grown as a potted or tubbed specimen reaching but a few feet in height. Branches that revert to the normal growth should be removed. They are all slow growers.

Euphorbia neriifolia

HEDGE EUPHORBIA P T S

ZONE 4

Euphorbiaceae (Euphorbia Family)

'Variegata'

'Cristata'

Native to dry places from western India eastward to New Guinea, this euphorb will reach 20 feet in height. It is a moderate grower. It is characterized by its persistent, succulent leaves carried at branch tips. It has a low spreading canopy not well

Acalypha wilkesiana

suited to providing shade. It makes an excellent hedge, screen, or windbreak. It performs best in full sun in a well-drained soil, although it is not particular regarding soil type. There is a form with green and yellow leaf variegation. Hedge euphorbia is heat, drought, wind, and salt tolerant, an excellent candidate for both the xeriscape and the beach garden.

Euphorbia royleana

NORTH INDIAN SPURGE P T S

ZONE 4

Euphorbiaceae (Euphorbia Family)

This euphorb is native to dry places in northern India both in hot locations and in cooler areas of the Himalayas. It will tolerate considerable cold. A moderate grower, it will eventually reach 20 feet in height, with multiple branches tipped with large, persistent, succulent leaves. It makes an excellent hedge, screen, or small shade tree. It is drought, heat, wind, and salt tolerant, a good choice for the xeriscape as well as the coastal garden.

Ficus microcarpa var. *crassifolia*

WAX FIG, TAIWAN FIG

ZONE 3

Moraceae (Mulberry Family)

Reaching 12 feet in height, this sprawling Asian evergreen makes a good, tall, dense bank cover or, with pruning, a thick hedge, screen, or even a small tree. It has good salt, wind, heat, and drought tolerance and thrives both at the seashore and inland. Although not particular as to soil, it produces best in an open, well-drained soil. It will grow in sand. It is a fairly rapid grower.

Gardenia volkensii

BUSHVELD GARDENIA

ZONE 3

Rubiaceae (Coffee Family)

Bearing an abundant crop of large, white, fragrant flowers, this African native will grow very slowly to 25 feet in height, so it is treated here as a large shrub. It must have full sun and is best in leeward coastal locations where there is copious sun and heat. Flowers are followed by large, decorative fruit that may be used in arrangements. A well-drained soil will produce good growth. Use it as an unusual flowering accent or as a hedge, windbreak, or screen. Plant it where its fragrance can be appreciated.

Grewia occidentalis

LAVENDER STARFLOWER

ZONE 3

Tiliaceae (Linden Family)

Originating in South Africa, this species will reach 10 feet in height, producing a dense, bushy growth good in mass plantings or espaliered. It flowers much of the year, more heavily at cooler elevations. It prefers full sun but will perform well in light shade. It has moderate tolerance of drought and is a moderate grower.

Hibiscus rosa-sinensis

RED HIBISCUS, COMMON HIBISCUS, CHINESE HIBISCUS

ZONE 2

Malvaceae (Hibiscus Family)

An evergreen Asian plant, hibiscus is one of the most favored plants in the landscape. The common red hibiscus will grow to 15 feet in height, flowering all year. Flowers are bright red and last only one day. It will do satisfactorily in any well-drained soil in full sun or light shade. It tolerates heat and wind and shows moderate tolerance to drought once well established. It makes a fine hedge or screen. Its many hybrids vary in height from low shrubs growing 4 or 5 feet in height and spreading equally to 20 feet in height. Colors vary greatly from white, pink, yellow, orange, and red to bicolors, even coffee and tan with a red eye. All prefer full sun, a good, well-drained soil, and regular feeding. They are heat and wind tolerant but have only moderate tolerance to drought and salt. Flowers are used in arrangements, sometimes skewered on a coconut leaf rib or placed in a shallow bowl. Water is not required. They are also a favored hair ornament. They are rapid growers.

Hibiscus schizopetalus

CORAL HIBISCUS, FRINGED HIBISCUS, ALOALO KOʻAKOʻA

ZONE 2

Malvaceae (Hibiscus Family)

Reaching 15 feet in height, this tropical East African, weeping hibiscus bears its unusual flowers all year. A number of hybrids are available showing the parental fringed flower petals. While the species is sprawling, requiring a wide planting space, its hybrids tend to be more columnar in form. All must have full sun in a well-drained area. They are wind tolerant, moderately drought tolerant, but not salt tolerant. They make beautiful specimen plantings or excellent hedges and screens. Flowers last only one day, but new flowers are produced daily. Flowers are used in arrangements, frequently skewered on coconut leaf ribs. They are rapid growers.

Holmskioldia sanguinea

CUP AND SAUCER, CHINESE HAT PLANT, PARASOL FLOWER

ZONE 3

Verbenaceae (Verbena Family)

'Citrina'

Originating in the foothills of the Indian Himalayas, this colorful, somewhat sprawling evergreen shrub will reach a height of 30 feet. It flowers much of the year. It may be used as a large hedge, screen, or color accent. Due to its sprawling habit of growth, it is not suited for tailoring into a small tree. Its flowers are used by the lei maker. The yellow cup and saucer (*Holmskioldia sanguinea* 'Citrina') provides the same landscape uses. There are several other color forms in the red to yellow range. All fare best in full sun in a well-drained soil. They are wind and heat tolerant, moderately tolerant of drought, but not tolerant of salt. They are rapid growers.

Kalanchoe hildebrandtii

HILDEBRANDT'S KALANCHOE

ZONE 4

Crassulaceae (Stonecrop Family)

A large species, this Madagascan shrub or small tree will somewhat slowly grow to 15 feet in height. Due to its bright green leaves covered with a whitish pubescence, it provides an interesting color accent in the dry garden, especially at night with uplighting. It makes a fine hedge, screen, or windbreak. Plant it in a well-drained soil in full sun. It is wind, heat, and drought tolerant.

Lysiloma thornberi

FEATHER BUSH

ZONE 3

Fabaceae (Bean Family)

At its best in full sun in a well-drained soil, this species is native to dry parts of the southern United States. It has good heat, drought, and wind tolerance, and is moderately salt tolerant. It reaches 15 feet in height, with a rounded canopy of feathery foliage punctuated by white, fuzzy flowers. It makes a good screen, hedge, or fine-textured foliage background for larger, dark green foliaged specimens. It can readily be pruned into a small tree form. It is a slow grower.

Murraya paniculata

MOCK ORANGE, CHINESE BOX, ALAHE'E HAOLE **A**

ZONE 3 EVALUATE

Rutaceae (Citrus Family)

A rather slow growing, large shrub native to the area from India east to the Philippines, mock orange will reach 25 feet in height. It is one of the most popular plants for making a large hedge, screen, or windbreak or pruned into a small tree form. It can be clipped into a tight shape or left to its natural form of growth. It fares best in full sun but will tolerate light shade. A good, well-drained soil is preferred. Both its foliage and its fragrant white flowers are used in arrangements and in leis. It is wind tolerant but shows only moderate tolerance to salt and drought.

Nerium oleander

OLEANDER, COMMON OLEANDER, 'OLIANA **P + S +**

ZONE 4

Apocynaeae (Dogbane Family)

Native to the broad area from the Middle East to Japan, this evergreen plant will attain 30 feet in height. It produces a dense shrub from new canes rising from the ground each year. It flowers profusely on new wood in colors ranging from white, pink, salmon, yellow, and rose to red and dark red. There are double- and single-flowered forms, fragrant flowers, and variegated foliage, both green and white and green and yellow. Old canes should be removed to the ground to encourage new growth. Unwanted young canes should be pulled rather than cut to reduce basal growth. Oleander is highly heat and drought tolerant and has good wind and salt tolerance. It makes an excellent hedge, screen, windbreak, or colorful mass planting or color specimen accent and is readily pruned into a small tree. It is a moderate grower.

Pachypodium geayi

MADAGASCAR PALM **P S T**

ZONE 4

Apocynaceae (Dogbane Family)

Native to dry areas of southwest Madagascar, this slender treelike shrub shares its common name with a close relative, *P. lamerei,* also a Madagascan. This species grows rather slowly to 15 feet in height, bearing large white flowers. Its foliage is highly attractive. It must have a well-drained soil in a place in the xeriscape with full sun. It is heat, wind, and drought tolerant and may be used in groups as a startling accent against a high wall or placed singly as a vertical accent.

Pittosporum tobira

JAPANESE PITTOSPORUM, TOBIRA

ZONE 3 EVALUATE

Pittosporaceae (Pittosporum Famiy)

Originating from Japan and China, this large, tough evergreen shrub will slowly reach 25 feet in height. It does best in a good, well-drained soil in full sun or light shade. It has good wind, salt, and drought tolerance and makes a fine hedge, screen, or windbreak. It can be trained into a small tree shape. Branches may be used in arrangements. It fares better at cool elevations but also works well along the coast. It is useful both in the xeriscape and the beach garden but does not perform well in pure sand.

Plumeria alba

WHITE FRANGIPANI TREE **S**

ZONE 3

Apocynaeae (Dogbane Family)

Native to semidry parts of Central America and the West Indies, this plumeria provides a striking foliar accent in the xeriscape. Its long, dark green leaves may be as slender as 1/4 to 5/8 of an inch wide, producing a startling contrast to other foliage. It has white, fragrant flowers. It is heat and wind tolerant but only moderately drought tolerant. Plant it in full sun in an open, well-drained soil. It will rather slowly reach 15 feet in height.

Plumeria stenopetala

HISPANIOLA PLUMERIA S

ZONE 4

Apocynaceae (Dogbane Family)

A rare plumeria species from an arid zone of Hispaniola, this plant is heat, wind, and drought tolerant and will slowly reach 18 feet in height. It is very shrubby, with wide-spreading branches. Its white flowers are plentiful and fragrant, with narrow petals tending to twist slightly. In the xeriscape it may be used as a hedge, tall screen (although it is briefly deciduous), bank cover, or accent.

Plumeria stenophylla

DOMINICA PLUMERIA S

ZONE 4

Apocynaceae (Dogbane Family)

An unusual deciduous plumeria from hot, dry locations in Dominica, this species will reach 18 feet in height, with wide-spreading branches. Flowers are white, fragrant, and numerous. It is heat, wind, and drought tolerant, performing best in full sun in a well-drained soil. In the xeriscape it finds use as a hedge or screen or as a foliar accent. It is rather slow growing.

Polyscias guilfoylei

PANAX, COMMON PANAX

ZONE 2

Araliaceae (Panax Family)

'Crispa'

A fast growing, evergreen, fastigiate shrub from southern Polynesia, panax is one of the most common hedge plants. It will reach 20 feet in height and is the perfect hedge or screen in narrow places or along a boundary where limited space is a factor. Plant it in full sun in any well-drained soil.

There is a form with nonvariegated foliage. It has moderate salt, wind, and drought tolerance. Many cultivars are available.

Portulacaria afra

ELEPHANT'S FOOD

ZONE 4

Portulacariaceae (Portulaca Family)

'Variegata'

A South African shrub reaching 12 feet in height, this species is heat, drought, salt, and wind tolerant and fire retardant. It presents a rather open form, with long, tapering stems. Plant it in full sun in any well-drained soil. There are slightly smaller variegated forms and one with larger leaves. Of these, *P. afra* 'Variegata', with yellow green leaves, is popular. Use it as an unclipped hedge or screen or a windbreak for more delicate low plantings. It is a rather slow grower.

Schefflera arboricola

DWARF BRASSAIA

ZONE 2 EVALUATE

Araliaceae (Panax Family)

'Emerald Ripple'

Actually an evergreen, bushy vine from Taiwan growing to 20 feet or more, this tough plant is usually seen as a controlled shrub and is so treated here. It does best in full sun but will perform satisfactorily in light shade in a well-drained soil. It has good wind and moderate drought tolerance and makes a good hedge, windbreak, espalier, or containerized specimen for a windy lanai. *S. arboricola* 'Variegata' is one of numerous variegated leaf forms available. It is a rapid grower.

Senna polyphylla

DESERT CASSIA

ZONE 3

Fabaceae (Bean Family)

Hot, dry areas of the Virgin Islands and Puerto Rico are the native habitat of this colorful species. Reaching up to 15 feet in height and developing a dense canopy, this shrub is heat, wind, and drought tolerant. It displays abundant golden yellow flowers almost year-round but especially during hot summer months. It requires full sun and a well-drained soil. Use it as a striking color accent, screen, or hedge or prune it into a small tree shape. It is a moderate grower.

Tabernaemontana divaricata

CREPE JASMINE, PAPER GARDENIA

ZONE 2

Apocynaceae (Dogbane Family)

'Fiore Pleno'

'Fiore Pleno'

This fast growing evergreen shrub from northern India and eastward to northern Thailand and southwestern China may reach 15 feet in height. Growing well in sun or shade, it prefers a rich soil with good drainage. Its plentiful white flowers have a yellow throat and are fragrant at night. Its wood is also fragrant and may be used for incense. It is useful as a hedge or screen, a background planting, or it may be pruned into a small tree. It shows moderate drought and wind tolerance but has no salt tolerance. There is a double form, the 'Butterfly Gardenia' (*T. divaricata* 'Flore Pleno').

Tamarix chinensis

CHINESE TAMARISK

ZONE 2

Tamaricaceae (Tamarisk Family)

A large shrub growing to 15 feet in height, this species is native to temperate areas of China, Tibet, Korea, and Japan. It does best in an open soil in full sun. It shows only moderate drought tolerance but thrives in coastal gardens, where it displays good tolerance of salt air. It can be used as a hedge, a screening plant, or a fine foliar accent among plants with dark green leaves. With uplighting, it makes a spectacular accent in the night garden. It is a slow grower.

Thunbergia erecta

BUSH THUNBERGIA, KING'S MANTLE

ZONE 2

Acanthaceae (Acanthus Family)

'Alba'

This evergreen African species grows slowly to 15 feet in height, bearing an abundance of blue purple flowers almost year-round. It is only moderately tolerant of heat, wind, drought, and salt air. It makes a striking color accent or can be planted as a hedge or screen or in a tall, mixed flowering border. Pruning is necessary to produce a small tree shape. There is also a white flowering form, *Thunbergia erecta* 'Alba'.

Vitex trifolia 'Purpurea'

PURPLE VITEX, ARABIAN LILAC

ZONE 4

Verbenaceae (Verbena Family)

Native to hot, dry areas of the Middle East to Australia, this tough shrub is wind, drought, and salt tolerant. It will grow to 25 feet in height and may be shaped into a tree form. It grows rather rapidly, developing a dense crown of somewhat fuzzy, gray-green leaves that are purple on their undersides. Use this vitex as a hedge, screen, low windbreak, or, as noted, pruned into a small tree. It is good at the beach and thrives in sand, but it does even better in soil.

Wilkesia gymnoxiphium

ILIAU

ZONE 3

Asteraceae (Sunflower Family)

This attractive endemic may reach 15 feet in height, usually seen with a single stem but occasionally multistemmed. In spring a terminal spike of cream-colored flowers provides both a color and textural accent. After flowering, the plant dies, but not until it produces abundant seeds. Use *iliau* singly or in a grove for an unusual garden effect. Its native habitat is the vicinity of Waimea Canyon, Kaua'i, on dry ridges or open forests. Under cultivation, however, it performs well at sea level and will flower and set seed if given an open, well-drained soil and full sun. It is a moderate grower.

Yucca aloifolia

SPANISH BAYONET

ZONE 4

Agavaceae (Agave Family)

Native from Florida and other southern states to the West Indies and Mexico, this species is exceptionally heat, wind, drought, and salt tolerant and will slowly grow to 15 feet in height. The dark green foliage is very sharply tipped. Great caution must be taken in handling the plant. Plant it where it is out of reach and where people cannot possibly come in contact with it. It is a rich color and foliar accent in the xeriscape and in the beach garden. Its white flowers are fragrant and seen in late spring and summer. Use it as a visual focal point or as a screen.

SMALL TREES
15–30 FEET

TREE NOTES

We list three groups of trees: small trees 15 to 30 feet in height, medium trees growing to 50 feet in height, and large trees, whose mature height exceeds 50 feet. Of equal importance is the canopy of the tree at maturity. Each species is genetically programmed to produce its height and canopy spread. Immature trees will exhibit the same proportions. These factors may be significantly modified by the proximity of other trees or buildings, by strong, persistent winds, by poor soil, and by water and drainage deficiencies. At the end of the text for each tree species, we have indicated in parentheses the probable mature height and canopy: (height 60/canopy 75), for example, indicates that this species will probably attain 60 feet in height and produce a canopy 75 feet across. The number of years required to reach maturity is not specified, as such variables as soil, water, wind, and soil fertility present an impossible problem. We note only that growth may be rapid, moderate, or slow.

Acacia seyal var. *seyal*

RED ACACIA, SHITTAH TREE N T

ZONE 4

Fabaceae (Bean Family)

var. *seyal*

var. *fistula*

var. *fistula*

var. *fistula*

This unusual tree is native to hot, dry areas from Egypt south to Somalia, Kenya, and to areas of tropical West African sub-Saharan regions. A moderate grower, it will reach 30 feet in height, forming a rather flat-topped tree with complex branching. Bark is highly colored: branches are covered with a bright reddish brown powder. In the nightscape it provides a dramatic color accent. Its abundant flowers are bright yellow. Parts of the tree produce gum arabic, a substance of many uses since ancient times, such as medicines and tannic acid for tanning. Shittim wood is said to have been used in Egypt to make coffins. It is heat, wind, and drought tolerant. References indicate its ability to tolerate as little as 20 inches of rain annually and to withstand a dry season of six to eight months. It provides a unique color accent in the xeriscape as well as light shade, which permits the growth of other species beneath its canopy. Seedling growth is very bushy. Thoughtful pruning is needed to expedite upward growth toward achieving a tree form. It is worth the trouble. A similar variety, *A. seyal* var. *fistula,* the green acacia, occurs in more southerly areas, although their ranges may overlap. Its bark is whitish to greenish yellow. It has the same growth requirements and uses as the red acacia. Both have large thorns that are easily clipped off of branches possibly within contact by the gardener and visitors. Both require pruning to produce a proper tree shape. (height 30 feet/canopy 30 feet)

Anacardium occidentale

CASHEW NUT P NT N+

ZONE 4

Anacardiaceae (Cashew Family)

This tree is drought, wind, and salt tolerant and is native to tropical America. It is best known from dry areas of coastal northeastern Brazil. Its attractive foliage develops into a wide-spreading canopy reaching 30 feet in height. New foliage is reddish in color. A very popular product, the cashew nut, comes from this species. Great care must be taken, however, to process the fruit according to directions. Fruit incorrectly processed is very toxic, although high in essential vitamins. Use protective gloves when handling the fruit and, probably, all parts of the tree, as juice and sap may cause burns on susceptible skin. Many gardeners discard the nut itself for fear of imperfect processing but eat the delicious, nontoxic red base of the fruit. Plant cashew in full sun in a well-drained soil. It is a moderately slow grower. It is recommended for both the xeriscape and the coastal garden. (height 30 feet/canopy 50 feet)

Annona muricata

SOURSOP **P**

ZONE 3

Annonaceae (Custard Apple Family)

The prized soursop tree, growing rather rapidly to 25 feet in height, is famous for its large fruit, containing a mass of juicy threads that are eaten raw or made into juice, ice cream, or sorbet. The fruit is high in vitamins B and C, and when immature it can be cooked with coconut milk. Native to the West Indies and Central America, it is now planted throughout the tropics. In Southeast Asia it is also valued for its medicinal qualities: a concoction of its leaves is said to induce sleep and have other medicinal and insecticidal properties. Seeds contain a toxin and may cause eye irritation. Its compact growth and thick, dark green leaves make it a good windbreak, screen, or an accent specimen. It is drought and wind tolerant and moderately salt tolerant and performs best in full sun in a well-drained soil. (height 25 feet/canopy 25 feet)

Bauhinia hookeri

ALIBANGBANG **N**

ZONE 3

Fabaceae (Bean Family)

Apparently this Australian tree has long been locally confused with a similar species native to the Philippines, and thus the Filipino name "alibangbang" is commonly applied to this very ornamental, slightly weeping tree. It will reach 25 feet in height and is a moderate grower. Basal suckers on young trees must be cut away. It has moderate salt and drought tolerance and good wind tolerance. Alibangbang has been used as a street tree and as an accent in the garden, where its almost everblooming flowers may be enjoyed. It fares best in full sun in a well-drained soil. (height 25 feet/canopy 25 feet)

Bauhinia tomentosa

YELLOW BAUHINIA **N**

ZONE 3

Fabaceae (Bean Family)

Growing to a height of 30 feet, this bauhinia is moderately drought and wind tolerant but not salt tolerant. Plant it in full sun in a well-drained soil. It flowers much of the year. Yellow bauhina is useful in the xeriscape as a color accent or tall screen. It has been used as a street tree. It is native to the wide geographical range from tropical Africa eastward to China. It is a moderate grower. (height 30 feet/canopy 30 feet).

Bolusanthus speciosus

RHODESIAN WISTERIA TREE N

ZONE 3

Fabaceae (Bean Family)

This is a beautiful, graceful tree that may grow to 30 feet in height. Flowers borne in spring on pendant racemes are a strong blue violet. Foliage is gray green, somewhat pendant. Plant it in a well-drained soil. It is fairly drought and wind tolerant. The Rhodesian wisteria tree is native to dry areas of Angola, South Africa, Rhodesia, and north into Mozambique. It will flower more heavily in cool, dry upland areas. Use it as an unusual color accent or where a light canopy is desired to permit undergrowth. It is a rather slow grower. (height 30 feet/canopy 20 feet)

Bursera microphylla

ELEPHANT TREE, COPAL, TOROTE COLORADO

ZONE 4

Burseraceae (Bursera Family)

Native to arid areas in Mexico, Baja California, and north into Arizona, the elephant tree grows slowly to 16 feet in height. It has a dense, wide-spreading crown with aromatic foliage reminiscent of camphor. It has a succulent, thick trunk with peeling bark, exposing colorful bark underneath. Its small white flowers are followed by bright red fruit. In the xeriscape use it as an accent or background plant or in the rockery. It is heat, wind, and drought tolerant. It requires full sun and an open, well-drained soil. (height 16 feet/canopy 25 feet)

Caesalpinia coriaria

DIVI DIVI N

ZONE 3

Fabaceae (Bean Family)

Growing to 30 feet in height, this tree originates from dry places in the Antilles, northern South America, and southern Mexico. It is spineless. Whitish or yellowish flowers are borne profusely in spring but are not conspicuous. It displays moderate tolerance to drought, heat, and wind and makes a useful shade tree or screen. It performs best in a light, well-drained soil. Divi divi has medicinal properties. It is a moderate grower. (height 30 feet/canopy 60 feet)

Caesalpinia pulcherrima

DWARF POINCIANA, PRIDE OF BARBADOS, 'OHAI ALI'I P T N

ZONE 4

Fabaceae (Bean Family)

forma *flava*

'Pride of Barbados'

'Compton'

forma *flava*

A small West Indian tree reaching 15 feet in height, this species flowers much of the year, bearing red or salmon-colored blooms. There is a yellow form, yellow dwarf poinciana (*C. pulcherrima* forma *flava*). All have good heat, salt, and drought tolerance. It flowers best if planted in full sun and is not particular as to soil as long as it is well drained. Use it as a bright color accent, hedge, or screen. Plant it where its flowers may be picked for lei making. It is a moderate grower. (height 15 feet/canopy 25 feet)

Callistemon citrinus

RED BOTTLEBRUSH, CRIMSON BOTTLEBRUSH

ZONE 2

Myrtaceae (Eucalyptus Family)

Bright red, bottle brush–shaped inflorescences up to 2.75 inches in length appear all year on the graceful weeping branches of this versatile Australian evergreen tree. New growth is colorful, showing pinkish or reddish leaves somewhat silky-hairy. It may reach 25 feet in height. It grows readily in a well-drained soil in full sun. It is drought and wind tolerant with moderate tolerance to salt. It makes a good shade tree, color accent, or light screen. It is a moderate grower. (height 25 feet/canopy 20 feet)

Callistemon rigidus

STIFF BOTTLEBRUSH

ZONE 2

Myrtaceae (Eucalyptus Family)

This Australian native will reach 20 feet in height, displaying a rather rigid, angular form. Its flowers are bright red and appear throughout the year. It has moderate drought, wind, and salt tolerance. Plant it a well-drained soil. The stiff bottlebrush finds good use as a colorful accent tree in the landscape and is a moderate grower. (height 20 feet/canopy 15 feet).

Callistemon viminalis

WEEPING BOTTLEBRUSH

ZONE 2

Myrtaceae (Eucalyptus Family)

This bottlebrush may grow to 30 feet in height, and as the name suggests, it takes a weeping habit. Its flowers, bunched together like a 5-inch-long bottle brush, are red, appearing mostly in spring. Many cultivars have been selected, some of which are favored for street tree plantings in its native Australia. A well-drained soil is optimum, as is full sun. It has moderate heat, drought, wind, and salt tolerance and is a moderate grower. (height 30 feet/canopy 25 feet).

Cassia roxburghii

RED CASSIA, RED LABURNUM N

ZONE 3

Fabaceae (Bean Family)

Red cassia is native to hot, dry places in south India and Sri Lanka. In a well-drained soil in full sun, it will reach 30 feet in height, with a broad canopy. It is a moderate grower. Its showy red flowers appear in late September and November when other trees are not in flower. It shows a fair tolerance of wind and good tolerance of heat and drought. Use this fine tree in the xeriscape as a color accent and for its shade-producing, evergreen canopy. (height 30 feet/canopy 35 feet)

Chloroleucon tortum

BRAZIL RAIN TREE T PN

ZONE 4

Fabaceae (Bean Family)

A moderate grower, this handsome tree will reach 25 feet in height. It develops an interesting branching pattern and displays a smooth, beautiful bark. Its spines are encased by the bark as it matures. Avoid heavy pruning, which encourages spiny new growth. It is native to dry, coastal Brazilian forest where it grows in any well-drained soil, even on sand dunes. Give it a place in the xeriscape in full sun, where it will produce an accent specimen. The bark nightlighted is eye catching. It is wind, drought, and heat tolerant and will tolerate salt wind given some protection. (height 25 feet/canopy 35 feet)

Clusia rosea

AUTOGRAPH TREE

ZONE 4 EVALUATE

Clusiaceae (Clusia Family)

'Variegata'

This tough tree, reaching 30 feet or more in height, is found throughout the Caribbean. It is heat, wind, drought, and salt tolerant and will grow in almost any soil, rock, or sand. Seeds germinate readily in both wet and dry conditions on stone walls, tree crotches, and road cuts, and the tree is considered invasive in many areas. It is indestructible in the xeriscape or the beach garden. It makes a fine screen, windbreak, or shade tree. The name "autograph tree" is derived from the fact that if a leaf is scratched, it leaves a persistent white mark. Leaves have been used as dinner table place cards and as a substitute for playing cards. A variegated form is available. (height 30 feet+/canopy 60 feet).

Cordia lutea

PERUVIAN CORDIA

ZONE 4

Boraginaceae (Borage Family)

A tough, small tree from dry areas of Peru, Ecuador, the Galapagos Islands, and the Marquesas, this species will reach 25 feet in height. Heavy pruning is needed to shape it into a tree form. It is drought and wind tolerant and has a fair tolerance of salt air. Its bright, clear, yellow flowers appear most of the year. Use it in a problem area in the xeriscape as a color accent, hedge, or screen. It prefers a light, well-drained soil and full sun. It is a slow grower. (height 25 feet/canopy 25 feet)

Cordia sebestena

KOU HAOLE, GEIGER TREE

ZONE 3

Boraginaceae (Borage Family)

Originating from tropical America and islands of the Caribbean, this colorful tree grows slowly to 25 feet in height. It bears clusters of brilliant orange red flowers much of the year. It grows well in any soil with good drainage. It prefers full sun. Kou haole has good wind, salt air, and drought tolerance. (height 25 feet/canopy 30 feet)

Crescentia cujete

CALABASH TREE

ZONE 3

Bignoniaceae (Catalpa Family)

Displaying a rather spreading canopy, this tree from Mexico, northern Central America, and the Caribbean islands will grow to 30 feet in height. It is wind and drought resistant and moderately salt tolerant. Its common name refers to the large, spherical fruit borne directly on its branches and trunk. These may be dried, hollowed, and used as musical instruments. Its corky bark is a good place for planting small epiphytes such as drought-tolerant bromeliads. It will grow satisfactorily in any well-drained soil and in full sun. It is a slow grower. (height 30 feet/canopy 35 feet)

Delonix decaryi

WHITE POINCIANA

ZONE 4

Fabaceae (Bean Family)

A rather slow grower eventually reaching 30 feet in height, this unique species from the dry scrub forests of western and southwestern Madagascar develops a somewhat swollen or cigar-shaped trunk. Its gray bark is smooth, flaking to reveal greenish or red brown bark. Flowering during the warm months, it produces an abundant crop of fragrant white flowers margined with apricot yellow. The showy upper petal sports a lemon yellow blotch. In its native areas, its resin produces a glue. Immature seeds are reputedly edible, while the trunk is used to make canoes. It is wind, heat, and drought tolerant and shows moderate tolerance of salt air. Use this unusual species as a special color accent in the xeriscape for both its flower and bark color. Its rather columnar shape does not provide significant shade. It produces best in full sun and when given an open, well-drained soil. (height 30 feet/canopy 15 feet)

Diphysa americana

GUACHIPILIN, CUACHEPIL

ZONE 3

Fabaceae (Bean Family)

Growing to 30 feet in height, this moderately slow growing Mexican and Central American tree produces a fine, almost lacy foliage, contrasting with its bright yellow blossoms. Plant it in any well-drained soil in full sun. It is wind tolerant and moderately drought tolerant. Its bark is rough and suitable for growing epiphytes such as small species of drought-tolerant orchids and bromeliads. Use it for its light shade, which permits the growing of understory species, or as a color accent. Note that one reference credits this species with growing to a height of 60 feet under optimum conditions. A local forty-year-old specimen, however, has reached only 25 feet in height. It is a moderate grower. (height 30 feet/canopy 35 feet)

Dracaena cinnabari

SOCOTRA DRAGON TREE

ZONE 4

Agavaceae (Agave Family)

This is a rare species from the island of Socotra. Although reaching 30 feet in height with many branches, it is so slow growing that we are considering it as a small tree. It eventually forms a dense, broad canopy of foliage. It is heat, drought, and wind tolerant. Full sun and very good drainage are necessary. Its foliage color and shape makes this species an interesting foliar accent. (height 30 feet/canopy 40 feet)

Dracaena draco

CANARY ISLAND DRAGON TREE

ZONE 3

Agavaceae (Agave Family)

The Canary Islands are the home of this popular 30-foot, branching tree with terminal clusters of blue gray leaves. It is a moderate grower in full sun and in a well-drained soil. It is wind, heat, and drought tolerant. Sprays of white flowers are followed by bright orange fruit. Leaf bases are orange and are prized by arrangers. Use it as an attractive accent and where its bluish gray foliage may be a foil for dark greens. (height 30 feet/canopy 35 feet)

Dracaena marginata

MADAGASCAR DRAGON TREE, MONEY TREE

ZONE 3

Agavaceae (Agave Family)

'Tricolor'

'Colorama'

'Majenta'

As its common name implies, this is a tree from Madagascar reaching 25 feet in height. It is multibranched, forming a wide canopy of slender mahogany-edged leaves. It produces white flowers in winter. Plant it in full sun in a well-drained soil. It is tough and withstands considerable abuse. It rewards the good gardener with a handsome specimen tree for the xeriscape. It is heat, wind, and drought tolerant. Several foliage color forms are available with white, green, and pink striping. The lighter foliage variations are dramatic color accents given uplighting in the nightscape. All are moderate growers. (height 25 feet/canopy 35 feet)

Dracaena marginata 'Grandifolia'

LARGE LEAF MADAGASCAR DRAGON TREE

ZONE 3

Agavaceae (Agave Family)

Cuttings of this large leaf dracaena were introduced from Tahiti, where it is seen, but not commonly, on the leeward side of the island and in the landscape of the old airport. It grows to 30 feet in height, producing multiple trunks topped by large rosettes of strongly mahogany-margined leaves. A local informant in Papeete claimed it was brought in from Madagascar when it was under French rule. It appears to be a robust form of *Dracaena marginata,* possibly fitting within the confines of the species. We have created a cultivar name with the hope that eventually more taxonomic information will establish a better name. It is heat, wind, and drought tolerant. It can be used as an eye-catching accent, screen, large hedge, or windbreak. It is a rather slow grower. It performs best in full sun in a good, well-drained soil. (height 30 feet/canopy 35 feet)

Eriobotrya japonica

LOQUAT

ZONE 2 EVALUATE

Rosaceae (Rose Family)

Growing rather slowly to 30 feet in height, this small tree from China and Japan develops a dense canopy of large, attractive foliage. At cool elevations, small orange fruits appear in winter following flowering. Blooms are white and fragrant. It needs an open, well-drained soil and full sun. It is wind tolerant and moderately drought tolerant. In the garden it provides a foliar accent and may be used as a screen, windbreak, and for its shade. (height 30 feet/canopy 30 feet)

A NOTE ABOUT ERYTHRINAS

Erythrinas are excellent garden subjects, and many species are candidates for the xeriscape. In 2005, however, the erythrina gall wasp (*Quadristichus erythrinae*) found its way to Hawai'i, with devastating results. Until recently released biological controls have been established islandwide, the planting of erythrinas is discouraged. Some species appear to be resistant to the pest, and two of these are described here.

Erythrina latissima

BROAD LEAVED CORAL BEAN P T N

ZONE 3

Fabaceae (Bean Family)

Growing very slowly to 30 feet in height, this small tree is native to dry places from South Africa to Mozambique. The common name is derived from the large size of its leaves, which are up to a foot across. It bears brilliant red flowers in late winter or early spring, followed by decorative beans that are bright red with a black spot. Caution: erythrina seeds contain a toxin and should not be ingested. In using the seed to make leis or jewelry, protect the nose and mouth from dust resulting from drilling the seed. This coral bean has a deeply furrowed, corky bark that may be used for the display of small, drought-tolerant epiphytes such as orchids and bromeliads. Plant it in full sun in a well-drained soil. It is heat, wind, and drought tolerant. This species is resistant to the erythrina gall wasp. (height 30 feet/canopy 30 feet)

Erythrina madagascariensis

MADAGASCAR CORAL BEAN, VOMBARA P T N

ZONE 3

Fabaceae (Bean Family)

Endemic to western and northern Madagascar and the Comoros Islands, this species produces bright red flowers during warm months, following a brief deciduous period. It is a moderate grower, reaching 30 feet in height and producing a broad canopy. It is wind and heat tolerant and moderately drought tolerant. Use it as a bright color accent, as a large screening plant, and as a shade tree. It requires an open, well-drained soil and full sun. It is resistant to the erythrina gall wasp. (height 30 feet/canopy 40 feet)

Eugenia uniflora

SURINAM CHERRY

ZONE 2 H(HPWRA)

Myrtaceae (Eucalyptus Family)

Growing slowly to 30 feet in height, this Brazilian tree finds many uses in the landscape. Abundant small, white flowers are followed by edible fruit used to make drinks, jams, and jellies. Its small foliage is aromatic, its bark decorative. It may be used as a natural hedge, clipped into a tight form, or used as an accent. It grows best in full sun in an open, well-drained soil. It shows only moderate tolerance of drought. (height 30 feet/canopy 30 feet)

Euphorbia cotinifolia

RED SPURGE, HIERBA MALA P S

ZONE 2

Euphorbiaceae (Euphorbia Family)

Best grown in full sun in a well-drained soil, this small evergreen tree from Mexico to northern South America will attain 20 feet in height. Use it as an important color accent, hedge, or screen. It shows some drought and wind tolerance but little tolerance of salt. Flowers are inconspicuous. Its toxic sap has many medicinal applications in its native areas. With uplighting in the night garden, it produces a dramatic color accent. It is a moderate grower. (height 20 feet/canopy 25 feet)

Euphorbia punicea

JAMAICAN POINSETTIA P

ZONE 3

Euphorbiaceae (Euphorbia Family)

A rare endemic species from dry parts of Jamaica, this handsome plant is described both as a shrub reaching 10 feet in height and as a tree growing to 30 feet in height. The latter is no doubt true when optimal conditions are met. A modicum of judicious pruning will help to produce a tree shape. Brilliant red or pinkish orange bracts at the base of the flower are seen most of the year. It will not tolerate "wet feet" and must have a well-drained soil in full sun. It makes an excellent, bright color accent in the xeriscape and the beach garden or as a hedge or screen. It is wind, heat, salt, and drought tolerant. Use care in handling, as its caustic sap may produce an irritation on susceptible skin. Wash it off with alcohol, as it is not soluble in water. This is true with all euporbs. It is a moderate grower. (height 30 feet/canopy 30 feet)

Euphorbia tirucalli

PENCIL TREE P T +

ZONE 4

Euphorbiaceae (Euphorbia Family)

'Sticks on Fire'

A popular succulent native to arid areas of tropical southern Africa and eastward to India and Indonesia, this plant will reach 25 feet in height. Its sap is very caustic/poisonous, so great care should be exercised when handling it. Use alcohol to remove sap, as it is not water soluble. It must have full sun and grows in any well-drained soil. It is heat, wind, salt, and drought tolerant. Use it as a tall hedge or screen, windbreak, or as a vertical accent among dark-foliaged plants. Plant it where people contact, especially children, is unlikely. Watch for root suckering, which tends to spread the clump. The use of a root barrier may prevent this. It does not have spines. It is a moderate grower. Also available is a cultivar, 'Sticks on Fire', which produces bright red new growth. This is a colorful addition to the dry garden. (height 25 feet/canopy 35 feet)

Euphorbia xylophylloides

WOODY LEAF EUPHORBIA P

ZONE 4

Euphorbiaceae (Euphorbia Family)

This slow-growing succulent species from Madagascar will eventually reach 20 feet in height, with a rather columnar shape. It is wind, heat, and drought tolerant and prefers a well-drained soil in a place enjoying full sun. It makes a fine screen, hedge, or windbreak or as a textural foil for species with large, round leaves. (height 20 feet/canopy 15 feet)

Ficus carica

COMMON FIG, FIG TREE

ZONE 3

Moraceae (Mulberry Family)

Cultivated since ancient times for its edible fruit, eaten either raw or dried, the fig reaches 30 feet in height. It is a moderate grower. Native to rocky, hot dry areas of the Middle East, it is also valued for its shade. Two cultivars are recommended for local gardens: 'Excel', a white fig, and 'Kadota', with purple skin. The common fig is heat, wind, and drought tolerant. It figures in many legends and myths. It is credited with being mankind's first clothing, as its leaves were strategically placed to hide Adam and Eve's nakedness. Ancient Greeks credited Bacchus with the creation of the fig. Ancient Egyptians believed that the fig was the favorite food of the Blessed Dead. Mohammed is quoted as saying, "If I were to say that any fruit had come down from Paradise, I would say it of the Fig." (height 30 feet/canopy 35 feet)

Fouquieria macdougalii

MEXICAN TREE OCOTILLO T

ZONE 4

Fouquieriaceae (Fouquieria Family)

Native to arid parts of Sonoran Mexico, this shrubby plant will very slowly grow to 20 feet in height. It bears bright green foliage, which is shed under drought stress. Its bark is an attractive light green, while its plentiful panicles of tubular red flowers provide additional color for the xeriscape. They appear in spring and summer. It must have full sun and a well-drained soil. It is heat, wind, and drought tolerant and makes a good color accent tree, barrier hedge, or screen. (height 20 feet/canopy 30 feet)

Gardenia brighamii

NĀNŪ, NĀ'Ū, HAWAIIAN GARDENIA

ZONE 4

Rubiaceae (Coffee Family)

An endangered Hawaiian endemic, *nānū* is now found in only a few dry places in Hawai'i. It will grow slowly to 20 feet or more bearing white, fragrant blossoms that are sought by the lei maker. It has attractive, dark green, dense foliage. It is heat, wind, and drought tolerant. Plant *nānū* in an open, well-drained loam in full sun. It figures in several old Hawaiian accounts and legends. Plant it where the fragrance of its flowers can be appreciated and accessible to the lei maker. Use it as a large shrub or small tree. (height 20 feet/canopy 20 feet)

Gardenia latifolia

GOURRI CUP

ZONE 3

Rubiaceae (Coffee Family)

Native to hot, dry areas of southern India and Sri Lanka, this endangered gardenia will eventually reach 30 feet in height. It is a slow grower. Pruning will hasten its development into a small tree. Early growth forms a large, dense shrub with a rounded canopy bearing a large crop of fragrant, white flowers. Plant it in full sun. It is heat, wind, and drought tolerant. Use this gardenia for the fragrance of its flowers, as an accent screen, or hedge. The lei maker might find the blossoms useful. They are similar to the Hawaiian endemic gardenia *nānū* (*G. brighamii*). (height 30 feet/canopy 30 feet)

Gossypium barbadense

SEA ISLAND COTTON

ZONE 4

Malvaceae (Hibiscus Family)

Forming a small tree to 15 feet in height, this rapid-growing but short-lived species provides a good filler while the xeriscaper awaits the growth of permanent species. It is native to tropical America and is the cotton grown commercially in the United States, which provides an interesting garden subject. Sea island cotton has naturalized in very dry places in Hawai'i, showing excellent drought tolerance. It is also wind and heat tolerant and will hold up under moderate salt conditions. Grow it in full sun in any well-drained soil. (height 15 feet/canopy 15 feet)

Guaiacum guatemalense

GUATEMALAN LIGNUM VITAE

ZONE 4

Zygophyllaceae
(Lignum Vitae Family)

Similar to its two "cousins" below, this small, tough tree will grow slowly to 25 feet in height. It is heat and wind tolerant and very drought tolerant, an excellent choice for the xeriscape. It, too, bears blue flowers. Cultural requirements are the same for the other species. Some authorities consider this species to be the same as *Guaiacum sanctum* below. (height 25 feet/canopy 25 feet)

Guaiacum officinale

LIGNUM VITAE

ZONE 3

Zygophyllaceae
(Lignum Vitae Family)

Originating in dry parts of the Bahamas, Cuba, the West Indies, and northern South America, lignum vitae is a beautiful evergreen but slow-growing tree reaching 30 feet in height. It produces masses of lavender blue flowers in late spring and summer followed by clusters of orange fruit that persist into fall. Its bark is handsome: mature grayish brown bark peels away in patches, exposing young green bark. Lignum vitae has medicinal applications. Its very hard wood is valued for use in fashioning hard-use articles. Plant it where its flowers, fruit, and bark may be appreciated. It makes a good specimen tree or color focal point. It is heat, wind, and drought tolerant but less tolerant of salt. (height 30 feet/canopy 30 feet)

Guaiacum sanctum

HOLYWOOD LIGNUM VITAE

ZONE 3

Zygophyllaceae
(Lignum Vitae Family)

This evergreen, slow-growing species is native to dry parts of Central America, the Caribbean, and north to the Bahamas and Florida Keys. Its size, cultural requirements, and garden uses are similar to those of its more commonly seen cousin, lignum vitae. In spring it bears flowers of an intense, deep blue. It is heat and wind tolerant and moderately drought tolerant. (height 20 feet/canopy 25 feet)

Ilex vomitoria

YAUPON HOLLY

ZONE 2

Aquifoliaceae (Holly Family)

'Pendula'

'Pendula'

Slowly reaching 24 feet in height, this species is native to southeastern Virginia south to Florida and west to Texas. It is heat, wind, salt, and drought tolerant. It grows best in full sun and is not particular as to soil, thriving in both sandy or clay soils as well as acid to alkaline soils. It can be readily pruned into a small tree shape, providing excellent shade, or used as a hedge, windbreak, or screen. To enjoy its dense clusters of red fruit, it is necessary, as with other hollies, to plant both male and female flowering specimens, as the sexes are borne on separate trees. There are useful cultivars: *Ilex vomitoria* 'Pendula', or weeping yaupon holly, with pendant branches. It will reach 20 feet in height. *Ilex vomitoria* 'Aurea' produces yellow berries. There is also a dwarf form (see "Medium Shrubs"). (height 20 feet/canopy 15 feet)

Ipomoea arborescens

CASAHUATE, PALO DEL MUERTO P

ZONE 4

Convolvulaceae (Morning Glory Family)

Native to tropical, arid areas of Sonoran Mexico, this is a small tree reaching 25 feet in height. It wants full, hot sun and good drainage. It produces large white flowers on its arching branches during winter, when it may be partially deciduous. Its bark is smooth and white. Seeds are poisonous and contain hallucinogenic substances. Its Spanish name, palo del muerte, means "walking stick of the dead," no doubt in reference to its chemical constituents. It also has medicinal uses. Plant this tree for shade and for a bright accent. It is heat, wind, and drought tolerant. (height 25 feet/canopy 20 feet)

Jacquinia pungens

SAN JUANICO, CUDJOE WOOD T

ZONE 4

Theophrastaceae (Theophrastus Family)

Very slowly growing to 20 feet in height, this species from dry areas of tropical western Mexico produces masses of small orange flowers primarily in summer but also sporadically during the year. These are used in its native area to make leis that retain their color after drying. When submerged in water, they regain much of their freshness and are highly fragrant. Its abundant fruits are orange and spherical. Each leaf is tipped with a small spine, which makes this tough plant an excellent barrier in the xeriscape. It can also be used as a screen or hedge. Its slow growth probably indicates its use more as a shrub than a small tree. It is drought, heat, and wind tolerant. Plant it in full sun in a well-drained soil. (height 20 feet/canopy 20 feet)

Jatropha multifida

CORAL PLANT T

ZONE 3

Euphorbiaceae (Euphorbia Family)

This tropical American species will reach 20 feet in height. It is a moderate grower, thriving in full sun in a well-drained soil. Seeds are highly toxic and should be handled with care and kept away from children. Immature seeds should be removed. Coral plant is heat, wind, and drought tolerant and shows fair tolerance of salt air but not direct salt spray. Use this attractive species as a color accent easily pruned to a small tree shape or use it en masse as a colorful hedge or screen. (height 20 feet/canopy 15 feet)

Kokia drynarioides

KOKI'O, HAU HELE 'ULA

ZONE 3

Malvaceae (Hibiscus Family)

Reaching 25 feet in height, this endemic, endangered species found in areas of 1,000 feet in elevation, also grows and flowers at lower elevations given a very well-drained soil and full, hot sun. It bears large ornamental leaves and large, brilliant red flowers. It was traditionally cultivated by Hawaiians for its bark, which yields a dye, and for its flowers, which may be used by the lei maker. It is found only at Pu'uwa'awa'a and Hu'ehu'e in North Kona and is near extinction. Fortunately it is readily propagated from seed and does well under cultivation. It is drought and wind tolerant but shows no tolerance to salt. (height 25 feet/20 feet)

Lagerstroemia indica

CRAPE MYRTLE

ZONE 3

Lythraceae (Crape Myrtle Family)

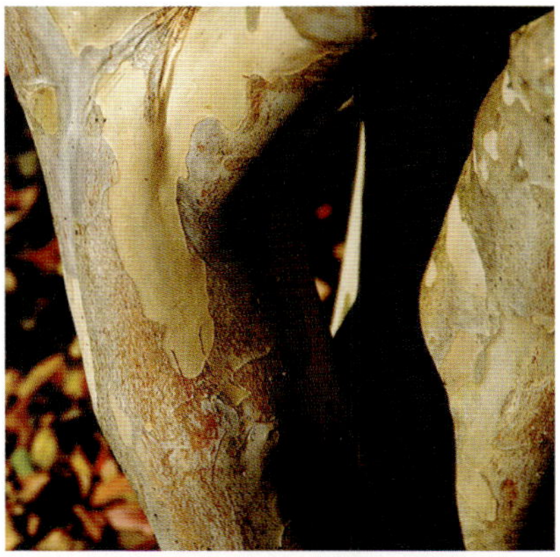

'Peppermint Lace'

Crape myrtle bears showy masses of flowers in summer. They range in color from white to pink and lavender purple. Its smooth bark is attractive, showing splotches of tan and gray. Attaining 30 feet in height, it makes an excellent shade tree or color accent specimen. It is native to China. It is wind tolerant and moderately drought tolerant once well established. Although not particular as to soil, crape myrtle produces best when given a friable, deep soil with good drainage. It is a moderate grower. Plant it in full sun. (height 30 feet/canopy 35 feet)

Lawsonia inermis

HENNA

ZONE 4

Lythraceae (Crape Myrtle Family)

Originating from North Africa to southern Asia, this is the source of the famous reddish orange dye used in ancient times by Egyptians and later by westerners and in the Middle East. Its leaves, crushed and mixed with an acid, produce the dye. Reaching 20 feet in height, it bears fragrant flowers of red or white that are noted in the Bible. It is heat, wind, and drought tolerant and may be used in the xeriscape as a small tree or accent or planted en masse for a good screen or windbreak. (height 20 feet/25 feet)

Lysiphyllum cunninghamii

KIMBERLEY BAUHINIA, BOHENIA TREE

ZONE 4

Fabaceae (Bean Family)

This small tree from dry parts of northern Australia will slowly grow to 30 feet in height. Flowers are pink with protruding red stamens and are borne liberally on its pendulous branches. Leaves vary in color from blue gray to gray green. Its bark contains an antibacterial substance used by Aborigines as an antiseptic wash. It has modern medicinal properties. A thick exudate of the bark, known as "toffee," is edible. It prefers an open, well-drained soil and does best in full sun. Use it as a color accent or shade tree. It is wind, heat, and drought tolerant. (height 30 feet/canopy 20 feet)

Majidea zanquebarica

ZANZIBAR SOAPBERRY

ZONE 3

Sapindaceae (Soapberry Family)

A highly variable species from East Africa and Madagascar, this soapberry is described in the literature as ranging from the size of a shrub to a timber tree and found in both dry and wet areas. Local experience has shown that it rather rapidly reaches 30 feet in height, showing moderate tolerance of drought. It forms a light canopy for shade and bears inconspicuous flowers, followed by highly colorful seed pods showing a bright red interior with dark, purplish blue, fuzzy seeds. The color persists after the pods dry. It is sought after by the lei maker as well as those who make wreaths and dry arrangements. Plant it in full sun in a well-drained soil. Use it for its shade and as a color accent. (height 30 feet/canopy 35 feet)

Malpighia emarginata

ACEROLA, BARBADOS CHERRY

ZONE 3

Malpighiaceae (Malpighia Family)

This tree, reaching 20 feet in height, is native to Central America. It performs best in a hot, dryish area in full sun in a well-drained soil. It produces small pink to purple flowers followed by bright red, tart fruit prized for its high vitamin C content. Up to three crops a year may be harvested. Superior cultivars have been developed, including 'Manoa Sweet', which bears nontart fruit, and 'B-15', with improved fruit and crop size. It may be used as a hedge, screen, or windbreak and for the edible garden. It shows moderate wind, heat, and drought tolerance but little salt tolerance. (height 20 feet/canopy 30 feet)

Melaleuca bracteata 'Revolution Gold'

BLACK TEA TREE, REVOLUTION GOLD

ZONE 2 H(HPWRA)

Myrtaceae (Eucalyptus Family)

This is a sport of a rather common Australian Melaleuca. It is valued for its bright yellow foliage. Growing moderately rapidly, it will reach 25 feet in height, developing a dense, broad canopy. Plant it in full sun in an open, well-drained soil. It provides a strong color accent and may be used as a windbreak or screen. It is wind tolerant but only moderately drought and salt tolerant. (height 25 feet/canopy 30 feet)

Euphorbia cotinifolia

Metrosideros polymorpha

ʻŌHIʻA LEHUA, ʻŌHIʻA, LEHUA

ZONE 3

Myrtaceae (Eucalyptus Family)

Usually thought of in terms of a large tree which may, under perfect conditions and much growing time, reach a height of 100 feet, the *ʻōhiʻa lehua* is of such slow growth that it is practical to consider it useful as a small tree to 30 feet in height. It provides rich color in the garden, with flowers ranging in color from pale yellow, strong yellow, orange, and apricot to red and dark red. Foliage is also variable from thick, dark green, fuzzy leaves to translucent leaves showing a reddish color. *Liko,* the unopened or newly opened leaf bud, is similarly variable, showing a wide range of color and texture. All are highly prized by the lei maker and figure heavily in Hawaiian legends and myths. Plant *ʻōhiʻa lehua* in full sun. A very well-drained soil is required. In the wild, the tree is found in very high rainfall areas as well as lowland arid places. Use it as a color accent, a focal point, windbreak, or hedge. Plant it where its flowers and foliage are convenient to harvest. It shows tolerance of heat, wind, drought, and salt. (height 30 feet/canopy 15 feet)

Morinda citrifolia

NONI, INDIAN MULBERRY

ZONE 4

Rubiaceae (Coffee Family)

Possibly native to southeast Asia, *noni* is now found in many tropical places. It has large, dark green foliage, developing into a round canopy accented by greenish white large fruit valued for its medicinal properties. Unhappily, fully ripe fruits emit a very unpleasant odor. Plant it downwind. It grows to 20 feet in height and is readily pruned into a small tree. It thrives in any open soil with good drainage as well as in lava. It is salt, wind, heat, and drought tolerant. Its dense growth makes it a candidate for use as a screening or windbreak plant. (height 20 feet/canopy 20 feet)

Moringa oleifera

HORSERADISH TREE, BEN OIL TREE, KALAMUNGAY

ZONE 4

Moringaceae (Drumstick Tree Family)

Now grown widely throughout the tropics, this small tree is native to dry areas of India and Arabia. It will grow rapidly to 30 feet in height in full sun and a well-drained soil. All parts of the tree furnish useful products: grated roots provide a horseradish substitute, and young, tender leaves, flowers, and immature seed pods are eaten as greens in salads, stir fries, and vegetable curries. Its leaves are high in protein and vitamins A and C. Ben oil is pressed from its seeds, which contain about 35 percent of the oil, which is used in cooking and for salads and to make soap. It does not turn rancid. It burns without smoke. It has been used as a special lubricant. In parts of Africa, branches are used as a charm against witchcraft. It is heat, wind, and drought tolerant. When harvesting leaves, be careful to respect the overall shape of the tree. Prune only the branch tips and refrain from massive cuts. In addition to its culinary uses, this lacy-foliaged tree provides a fine light shade and a bright green color to the xeriscape. It is basically recommended here as an excellent edible garden plant and not for the nonusing gardener. (height 30 feet/canopy 25 feet)

Munroidendron racemosum

MUNROIDENDRON

ZONE 3

Araliaceae (Ginseng Family)

Modereately drought, wind, and salt tolerant, this rare endemic, endangered species is readily grown under cultivation, reaching about 25 feet in height. It is a fairly rapid grower. Foliage is ornamental as is its unusual pendant inflorescence. Plant it in good, well-drained soil in full sun. This makes a fine accent or specimen plant. Munroidendron is endemic to Kaua‘i. (height 25 feet/canopy 20 feet)

Myoporum sandwicense

BASTARD SANDALWOOD, NAIO

ZONE 4

Myoporaceae (Naio Family)

This moderately slow growing, extremely tough native species reaches 30 feet in height and is best in full sun in a well-drained soil. It is wind, heat, drought, and salt tolerant. Use it as an accent tree, hedge, screen, or windbreak. There is a low, creeping form that has the same characteristics. It is listed under “Small Shrubs.” *Naio* wood, quite fragrant, was once tried as a substitute for sandalwood during the last days of the trade with China, when sandalwood was becoming scarce. The Chinese found it unacceptable. In traditional times, Hawaiians used its tough wood for framing houses. It is an excellent choice for both the xeriscape and the beach garden. (height 30 feet/30 feet)

Nolina recurvata

PONY TAIL, BEAUCARNEA

ZONE 4

Agavaceae (Agave Family)

'Variegata'

Deserts of northern Mexico are the native home of this popular woody succulent. A slow grower, it will reach 30 feet in height planted in full sun in a well-drained soil. It will develop multiple trunks and a broad base. It is drought and wind tolerant and moderately salt tolerant. Pony tail is a popular plant for a large pot or tub and will take any amount of abuse, shade, and neglect. In the garden, it is a striking accent. Flower arrangers prize its dried leaves. A variegated form is available. (height 30 feet/ canopy 25 feet)

Pachypodium lamerei

MADAGASCAR PALM P S

ZONE 4

Apocynaceae (Dogbane Family)

This is a small, branching, treelike succulent native to very dry areas of Madagascar. It will reach 15 feet (one reference indicates 20 feet) in height, each branch tipped with clusters of attractive white, fragrant flowers. Unfortunately, perhaps, the branches are all heavily armed with spines and care is needed in handling. It is, of course, not a palm. It requires full sun and a well-drained soil. It is wind, heat, and drought tolerant. Use it as a strong accent in the xeriscape, planted where its flowers can be appreciated but not where people are likely to brush against it. A similar species, *P. geayi,* also known as Madagascar palm, has attractive narrow, grayish leaves. It also bears white flowers. Both are rather slow growers. (height 15 feet/canopy 10 feet)

Parmentiera aculeata

CUAJILOTE, GUACHILOTE T

ZONE 3

Bignoniaceae (Catalpa Famly)

Growing rather slowly to 30 feet in height, this species originates in the area from tropical Mexico south to Guatemala. Its green and mauve flowers are borne directly on its trunk and larger branches. These are followed by sweet, yellow fruit eaten both cooked and raw. It needs full sun and a well-drained soil. It tolerates heat, wind, and moderate drought. Its rough bark makes a good home for drought-tolerant epiphytes such as small bromeliads. When pruning, take care to avoid its small spines located at leaf bases. Use this plant for its shade and its fruit, and as a conversation piece. (height 30 feet/ canopy 35 feet)

Pereskia grandifolia

PERESKIA

ZONE 4

Cactaceae (Cactus Family)

This drought-tolerant Brazilian species has become a popular xeriscape subject, reaching as much as 16 feet in height. Although a cactus, it produces large, dark green, glossy leaves. Flowers, almost roselike, are clustered at branch tips and may vary from white to pink and dark pink. It does have spines at leaf bases. Pereskia is a moderately slow grower, performing best in full sun and a well-drained soil. It is heat, wind, and drought tolerant. Use it as a textural and color accent, screen, barrier planting, or windbreak, or carefully pruned into a small shade tree. (height 16 feet/canopy 15 feet)

Petrea arborea 'Alba'

WHITE TREE PETREA

ZONE 3

Verbenaceae (Verbena Family)

Native to dry parts of Venezuela, this attractive small tree will reach 25 feet in height, bearing plentiful, pendant sprays of white flowers in summer. The flowers may be used by the lei maker. It is wind and drought tolerant and requires full sun in an open soil with good drainage. Use it as a striking accent. There is also a blue purple–flowered form, but as yet it is not available in Hawai'i. It is a moderate grower. (height 25 feet/canopy 25 feet)

Pittosporum confertiflorum

HŌ'AWA

ZONE 3

Pittosporaceae (Pittosporum Family)

Found at low elevations in dry to moderately moist areas, this native Hawaiian endemic species will reach 25 feet in height, bearing white to cream-colored flowers that are fragrant at night. Plant it in full sun in an open soil with good drainage. It is a good shade tree and can be used as a screen or hedge. Plant it where its night fragrance may be enjoyed. It is moderately drought and wind tolerant but shows little tolerance to salt. It is a rather slow grower. (height 25 feet/canopy 25 feet)

Pittosporum flocculosum

HŌ'AWA

ZONE 3

Pittosporaceae (Pittosporum Family)

This rare endemic tree from O'ahu in both moist and semidry places will reach 30 feet in height. It is a moderate grower. It needs full sun and a well-drained soil. Use it in the garden where its night-fragrant white flowers may be enjoyed. It can be used for its shade or as a screen or hedge. It is wind tolerant and moderately drought tolerant but not salt tolerant. It is a rather slow grower. (height 30 feet/canopy 30 feet)

Pittosporum viridiflorum

CAPE PITTOSPORUM

ZONE 3 EVALUATE

Pittosporaceae (Pittosporum Family)

A small, tough tree reaching 25 feet in height, this species is heat, wind, and moderately drought tolerant. It develops a dense canopy with small, fragrant flowers. It is slow growing. Use it for shade or screening, or as a windbreak. It is native to the broad area from South Africa and Madagascar to India. (25 feet/canopy 25 feet)

Pleomele hawaiiensis

HALA PEPE

ZONE 4

Agavaceae (Agave Family)

Once plentiful, this *hala pepe* is now rare. It is still found in dry areas of Kona at elevations between 1,500 and 2,400 feet, but it grows satisfactorily under cultivation at lower elevations. Growing slowly to 20 feet in height with multiple branches, the species produces abundant flowers on a pendant stalk. These are prized by the lei maker. It is sacred to Laka, Goddess of Hula. Full sun and excellent drainage are necessary for its growth. It is heat and wind tolerant and moderately drought tolerant. Use it for its flowers or use its vertical form as an accent. (height 20 feet/canopy 15 feet)

Plumeria caracasana

CARACAS PLUMERIA

ZONE 4

Apocynaceae (Dogbane Family)

Described variously as an erect shrub or small tree to 30 feet in height, this deciduous plumeria is native to hot, dry coastal areas of Venezuela and the adjacent island of Margarita. It produces abundant white, fragrant flowers from early spring to early fall. Use it in the xeriscape for its flowers and as a seasonal change to produce interest in the garden design. Give it a well-drained soil. Plant it in full sun. It is highly drought, heat, and wind tolerant and probably shows a fair tolerance to salt air. (height 30 feet/canopy 30 feet)

Plumeria obtusa

BAHAMAS PLUMERIA **P S**

ZONE 4

Apocynaceae (Dogbane Family)

A number of years ago, this tree was introduced from the Bahamas. It was then called *Plumeria bahamensis,* but it is now merged with *Plumeria obtusa* (Singapore plumeria), a highly variable species. We are taking the liberty of listing it as the Bahamas plumeria because of its visual difference, which is significant for the gardener. It is a moderate grower, reaching 25 feet in height. Leaves are about a quarter the size of the

Singapore plumeria. Its abundant flowers are white and fragrant and slightly smaller than those of the Singapore plumeria. Native to coastal areas of the Bahamas, it is salt, wind, heat, and drought tolerant. It is a good candidate for both the xeriscape and the beach landscape. Full sun is required for optimal growth. (height 25 feet/canopy 20 feet)

Plumeria obtusa var. *sericifolia*

SILKY LEAF PLUMERIA P S

ZONE 3

Apocynaceae (Dogbane Family)

A highly useful species from dry, coastal areas of Dominica, this plumeria is characterized by the rather silky-to-the-touch underside of its leaf and by the pyramid-shaped lumps that develop on older branches. It reaches 30 feet in height with an equal spread and is a moderate grower given full sun and a well-drained soil. It is heat, wind, drought, and moderately salt tolerant, a good candidate for both the xeriscape and the coastal garden. Its 2-inch-wide, fragrant white flowers are borne abundantly through spring and summer. Its evergreen foliage is dark green. In addition to its shade-giving quality, it may be used for a windbreak or screen. (height 30 feet/canopy 30 feet)

Plumeria pudica

PUERTO RICAN FRANGIPANI

ZONE 3

Apocynaceae (Dogbane Family)

An evergreen plumeria species from the West Indies (originally described from Curaçao) growing to 20 feet in height, this plant bears clusters of unscented white flowers and a canopy of unusual leaves described as "spoon shaped" or lobed. It needs protection from wind but is drought and salt tolerance. Plant it in full sun in a well-drained soil. It makes a handsome accent or can be used to make a hedge or screen. Plant it where its flowers may be appreciated. (height 20 feet/canopy 15 feet).

Plumeria rubra forma *acutifolia*

BAJA PLUMERIA, CACALOSUCHII

ZONE 4

Apocynaceae (Dogbane Family)

This plumeria occurs in dry areas with a distinct monsoonal climate such as southern Baja California and parts of mainland Mexico and Central America. It grows rather slowly to 25 feet in height, displaying a somewhat columnar form and bearing masses of white, fragrant, fluffy flowers. It is heat, wind, and drought tolerant. Photos taken in habitat show it growing among columnar cacti and deciduous thorn scrub. In the xeriscape use it where space is limited for an accent, for its shade, as a hedge or screen, but always in full sun. Provide a well-drained soil. The flowers are not useful for lei making. (height 25 feet/canopy 15 feet)

Psydrax odorata

ALAHE'E

ZONE 4

Rubiaceae (Coffee Family)

A native Hawaiian evergreen species, this tree will reach 30 in height. It displays dark, shiny green foliage and bears fragrant white flower clusters prized by the lei maker. It is a moderately slow grower favoring full sun, a good, well-drained soil, and regular feeding. It is drought and wind tolerant and successfully withstands salt wind. Use it as a handsome specimen against a light background or as a hedge, screen, or windbreak. (height 30 feet/canopy 30 feet)

Pterocarpus rohrii

MEXICAN PTEROCARPUS N

ZONE 3

Fabaceae (Bean Family)

A highly variable species from Mexico south through Central America, Brazil, Peru, and Bolivia, this tree is found in very dry to very wet sites and described as growing from 20 feet to 90 feet in height. The local specimen has not reached 20 feet in height in three decades, and it is safe to assume that it is derived from the dry climate Mexican source. It is a beautiful small tree with a broad, rounded canopy bearing masses of yellow flowers in late spring. It is obviously a slow grower, preferring a well-drained soil in full sun. Use it as a shade tree or color accent specimen. (height 20 feet/canopy 30 feet)

Punica granatum

POMEGRANATE

ZONE 4

Punicaceae (Pomegranate Family)

Probably originating in Iran and nearby areas, the pomegranate has been grown since ancient times for its bright red flowers and fruit. Reaching 20 feet in height, it prefers a well-drained soil and is tolerant of alkaline soils, heat, wind, and drought but is not tolerant of salt air. It makes a fine accent or may be used to form a hedge. A modicum of pruning is needed to assist the pomegranate in taking a tree shape. There is a form with double flowers, and there are dwarf forms. These are listed elsewhere in the text. In its native areas, pomegranate has many medicinal uses. A popular beverage, grenadine, is made from the pulp, while the rind is the source of a dye for coloring Moroccan leather. The flower is the national floral emblem of Spain. Ancient Egyptians made wine from its fruit. It is a rather slow grower. (height 20 feet/canopy 25 feet)

Scaevola taccada

TREE NAUPAKA
ZONE 4
Goodeniaceae (Naupaka Family)

Botanically the same as the common beach *naupaka,* this is a form found on some South Pacific islands. It grows to 20 feet in height and is wind, drought, and salt tolerant. Plant it in full sun in a well-drained soil or even in sand. It makes an excellent windbreak, screen, or tall hedge in the xeriscape and beach garden. It is easily pruned into a small tree. It is a moderate grower. (height 20 feet/canopy 30 feet).

Senna surattensis

KOLOMONA, KALAMONA, SCRAMBLED EGGS
ZONE 4 EVALUATE
Fabaceae (Bean Family)

Probably native to dry places in Southeast Asia, Australia, and Polynesia, this tough small tree will reach 20 feet in height. It forms a rather dense canopy with many yellow flowers almost all year. They are used by the lei maker. It is heat, wind, and drought tolerant and a good candidate for use in the xeriscape. It makes a good windbreak, screen, or color accent specimen and provides light shade. It is a moderate grower. (height 20 feet/canopy 20 feet)

Sesbania grandiflora

VEGETABLE HUMMINGBIRD, SESBAN, ʻOHAI KEʻOKEʻO N
ZONE 3
Fabaceae (Bean Family)

Native to tropical Asia, this rapid-growing small tree will reach 30 feet in height, bearing large white flowers. There is a rosy pink flowered form. Its canopy is light, consisting of attractive gray green leaves carried on a trunk with corky, furrowed bark. All parts of the tree are useful: flowers may be battered and fried, while tender young leaves and immature seed pods may be eaten raw in salads or added to soups and curries. Fiber, gums, a dye, and medicine are other derivatives. Full sun and a well-drained soil produce a robust specimen. It is heat, wind, and drought tolerant but shows only moderate tolerance of salt. Use it in the garden for its edible qualities, for its shade, and as a year-round flowering accent. (height 30 feet/canopy 30 feet)

Sophora chrysophylla

MĀMANE

ZONE 4

Fabaceae (Bean Family)

At low elevations, this endemic species is found in hot, dry situations where it is seen as a large shrub. Its most favorable environment, however, appears to be in dry, cold places such as the lee side of the saddle between Mauna Kea and Mauna Loa on Hawai'i, where it forms wide-canopied trees to 45 feet in height. It is found on all the main islands except Ni'ihau and Kaho'olawe. In flower, this is a spectacular color accent. The bright flowers are fashioned into leis. It may be used as a hedge, screen, or windbreak. Give it a well-drained soil in full sun. It is a slow grower. (height 10 feet/canopy 10 feet)

Tabebuia aurea

SILVER TRUMPET TREE

ZONE 3

Bignoniaceae (Catalpa Family)

A popular Brazilian tree reaching 30 feet in height, this is a bright accent in the xeriscape due to its silvery gray green foliage and abundant strong yellow flowers. Both leaf and flower color is variable, however, with some seedlings producing less silvery foliage and some with pale yellow flowers. Its bark is rough, furrowed, and attractive. Plant it in full sun in a soil with good drainage. It shows moderate drought, heat, and wind tolerance. Newly planted specimens should be well staked. It is a moderate grower. (height 30 feet/canopy 25 feet)

Tabebuia berteroi

HISPANIOLAN ROSY TRUMPET TREE

ZONE 3

Bignoniaceae (Catalpa Family)

Reaching up to 30 feet in height, this Hispaniolan (Caribbean) tree tends to be somewhat columnar in shape. It shows moderate heat, drought, and wind tolerance. Blossoms vary from almost white to light, rosy pink. Good drainage in any average soil and full sun are the ingredients for successful growth. It is useful as a shade tree, tall screen, windbreak, or as a color accent. It is a moderate grower. (height 30 feet/canopy 20 feet)

Thevetia peruviana

BE-STILL TREE P +

ZONE 4 H(HPWRA)

Apocynaceae (Dogbane Family)

Growing moderately rapidly, this tough tropical American species will reach 25 feet in height, producing a broad canopy. It bears bright yellow flowers all year around. Two cultivars are available: 'Alba', with white flowers, and 'Aurantiaca', with coral orange flowers. Care must be taken to avoid its sap, as all parts of the tree are toxic. Children are attracted to both its colorful blooms and rather unusual fruit. It performs well in almost any soil, as long as drainage is good, and it may be used for its shade, as a screen, hedge, windbreak, or as a color accent. It is wind, heat, and drought tolerant. It must have full sun. (height 25 feet/ canopy 35 feet)

Thevetia thevetioides

GIANT THEVETIA P +

ZONE 4

Apocynaceae (Dogbane Family)

Producing large, yellow, fragrant flowers throughout the year, this Mexican tree provides a striking color accent in the xeriscape. Its fine foliage is a foil for the bright flowers. It will grow moderately rapidly to 30 feet in height, forming a good shade tree, windbreak, or screen. Plant it in full sun in a well-drained soil. It is drought, heat, and wind tolerant but not tolerant of strong salt winds. All parts of the plant are toxic. Due to the attractiveness of both flowers and fruit, it is best not planted where it is accessible to small children. (30 feet/ canopy 35 feet)

Tournefortia argentea

BEACH HELIOTROPE, TAHINU

ZONE 3

Boraginacefae (Borage Family)

Native to beach areas of tropical Asia and the South Pacific, this rapidly growing tree may reach 25 feet in height. It is a fine choice for exposed beach gardens, showing excellent tolerance of heat, wind, and salt and moderate tolerance of drought. Planted in any well-drained soil or beach sand, it will quickly develop a broad canopy, providing shade, a windbreak, or a screen. Its bark is interesting, furrowed, and rough. Parts of the tree have medicinal uses, while its wood has been used for canoe building. It figures in Micronesian legends. Its large, succulent, translucent leaves glow when uplighted in the night garden. (height 25 feet/ canopy 40 feet)

Uncarina grandidieri

MOUSE TRAP TREE

ZONE 3

Pedaliaceae (Sesame Family)

This somewhat succulent tree, one of nine species endemic to Madagascar, will reach 25 feet in height. It is a moderate grower. It presents abundant yellow flowers during the year. Although heat tolerant, it is only moderately drought tolerant. Plant it in full sun in a friable soil with good drainage. Use it as an unusual color and foliar accent. A similar species, *Uncarina roeoesliana,* grows to 15 feet in height and can be used as a large shrub or small tree. (height 25 feet/canopy 20 feet)

Yucca guatemalensis

SPINELESS YUCCA, GIANT YUCCA

ZONE 4

Agavaceae (Agave Family)

'Marginata'

Native to dry parts of Mexico and Guatemala, this handsome tree will reach 30 feet in height. Foliage is long, narrow, and persistent even during periods of drought, and as the name implies, it is spineless. Large vertical flower spikes borne in spring and fall bear abundant, attractive white blooms. Plant it in full sun in a well-drained soil. It is drought, heat, and wind tolerant and shows moderate tolerance to salt wind. There is a variegated form showing leaves with white stripes. It is a moderate grower. (height 30 feet/canopy 20 feet)

MEDIUM TREES 30–50 FEET

ALOE SPECIES

For common names, those marked with a single asterisk (*) are listed by the Institute for Aloe Studies; two asterisks (**) indicate names found online; and three asterisks (***) indicate that no common name was found by the authors.

Aloe eminens

***

ZONE 4

Aloeaceae (Aloe Family)

One of tallest of the aloes, this rare giant from Somalia will slowly attain 50 feet in height. It branches and rebranches dichotomously, forming a broad canopy carrying branched spikes of red flowers. It must be given space, full sun, and a well-drained soil. It is heat, wind, and drought tolerant. It is a great accent specimen for the large xeriscape.
(height 50 feet/canopy 50 feet)

Bauhinia x *blakeana*

HONG KONG ORCHID TREE **N**

ZONE 2

Fabaceae (Bean Family)

Reaching 40 feet in height, this hybrid from China flowers throughout the year, more abundantly during the cool months. It does not produce seed pods. It makes a good color accent or shade tree and shows moderate tolerance of drought and wind but none to salt. It is used as a street tree and is a colorful accent in the landscape. Give it full sun and a well-drained soil. It is a moderate grower.
(height 40 feet/canopy 40 feet)

Bauhinia variegata

ORCHID TREE **N**

ZONE 2

Fabaceae (Bean Family)

'Candida'

Growing rapidly to 40 feet in height, this attractive tree is native to India and China. It develops a dense, round-headed canopy ornamented with flowers much of the year. Its somewhat haphazard growth habit requires careful pruning to prevent crossing branches. Its usual bounty crop of seed pods can be removed at the time of pruning, which is following flowering. Flower color ranges from white to light and dark purple. It has moderate tolerance of drought and wind but is not salt tolerant. Its wood is known as mountain ebony and is hard and heavy. Juvenile leaves and seed pods may be eaten. It may be used for its shade and for a color accent.
(height 40 feet/canopy 40 feet)

Butea frondosa

FLAME OF THE FOREST **NP**

ZONE 3

Fabaceae (Bean Family)

Native to hot, dry places in India and Sri Lanka, this deciduous member of the bean family will reach 40 feet in height. It bears clusters of spectacular, brilliant orange flowers when the tree is leafless. It tends toward a rather haphazard growth and must be pruned to maintain a good shape. Plant in full sun in a well-drained soil. Overwatering reduces flower production. It is drought, wind, and heat tolerant. In the xeriscape, it makes a dominant color accent. Plant it where its flowers may be seen close at hand. In a large property, it would make a startling high hedge or screen. In its native areas, the large leaves are used as plates, for wrapping paper, and as umbrellas. It is a rather slow grower.
(height 40 feet/canopy 30 feet)

Caesalpinia ferrea

BRAZILIAN IRONWOOD, LEOPARD TREE
ZONE 3
Fabaceae (Bean Family)

Growing moderately to 40 feet in height with an equal canopy spread, this Brazilian species produces fragrant yellow flowers in spring. Its common names derive from the hardness of its wood and the beautifully mottled colors of its light gray brown bark, which peels off to expose an almost white new bark. In full sun and a well-drained soil, this handsome tree shows resistance to heat and wind and moderate tolerance of drought. Use it in the xeriscape as a shade tree or a colorful accent, or in parks and roadway plantings.
(height 40 feet/canopy 40 feet)

Cassia fistula

YELLOW SHOWER, GOLDEN SHOWER, INDIAN LABURNUM N
ZONE 3
Fabaceae (Bean Family)

Prized for its long, pendant flower clusters borne from March to September, this Indian tree shows good drought tolerance and performs best in hot, dry leeward areas. It will grow to 40 feet in height. Its long, black seed pods are a litter problem and have an unpleasant odor if broken. It is a wonderful color accent and produces a good, light shade. Provide full sun and a well-drained soil.
(height 40 feet/canopy 40 feet)

Cassia grandis

PINK SHOWER, CORAL SHOWER **N**

ZONE 2

Fabaceae (Bean Family)

Native to tropical America, the pink shower tree flowers abundantly in early spring, forming an excellent color accent in the landscape. Growing to 50 feet in height, it provides good shade. Although thriving in a well-watered situation, it shows moderate tolerance to drought once well established. Its large pods present a seasonal litter problem. It is a moderate grower.
(height 50 feet/canopy 40 feet)

Cassia sieberiana

AFRICAN LABURNUM **N**

ZONE 3

Fabaceae (Bean Family)

From dry savannah areas of Senegal, Uganda, and Sudan, this colorful species will reach 50 feet in height. It carries a fairly open canopy, bearing abundant pendant racemes of bright yellow flowers all spring. It is a moderate grower and shows moderate tolerance of drought, heat, and wind. Its wood is used in cabinetry and carpentry. The tree has medicinal uses as well as provides material for a barbasco, a chemical for temporarily stunning fish so they may be easily caught.
(height 50 feet/canopy 40 feet)

Ceratonia siliqua

CAROB TREE **N**

ZONE 3

Fabaceae (Bean Family)

Native to dry parts of the eastern Mediterranean and Arabia, the carob tree is drought and heat tolerant and moderately salt tolerant. In temperate areas it will reach 40 feet in height, developing a dense, rounded canopy. Its plentiful fruit is used to produce alcohol and a substitute for chocolate and coffee. In Hawai'i, it does set fruit at lower elevations and reaches only about 30 feet in height. It can be used as a shade tree, a screen, or a windbreak. It is a slow grower.
(height 30 feet/canopy 30 feet)

Chorisia speciosa

FLOSS SILK TREE

ZONE 3

Bombacaceae (Bombax Family)

Native to dry areas of Brazil and Argentina, this deciduous tree will grow to 50 feet in height. It shows good tolerance to drought and moderate tolerance of salt. It must have a well-drained soil and full sun. Overwatering will reduce flower production. Flowers are spectacular, varying from pink to purplish rose and burgundy. Two named varieties are available: 'Los Angeles Beautiful' bears wine-red flowers; 'Majestic Beauty' displays rich pink blooms. Both must be grafted. It is an excellent color accent. Both are moderate growers. (height 50 feet/canopy 40 feet)

Coccoloba uvifera

SEA GRAPE

ZONE 3

Polygonaceae (Buckwheat Family)

A tropical American coastal tree, sea grape will reach 50 feet in height. Basal pruning will assist in producing a single-trunked tree; unpruned trees may produce multiple trunks, making sea grape a good screening or windbreak candidate. It is highly wind and salt tolerant and, once well established, quite drought tolerant. Its pendant, clustered fruit is made into a delicious jelly. If the fruit is not to be used, avoid fruit production and the attendant mess of overripe fruit on the ground by planting only male trees grown from cuttings or air layers. Sexes are conveniently borne on separate trees. It is a moderate grower. (height 50 feet/canopy 70 feet)

Colubrina oppositifolia

KAUILA

ZONE 3

Rhamnaceae (Buckthorn Family)

Rather slowly reaching 40 feet in height, this endangered endemic species is found in dry to mesic forest in the Wai'anae Range on O'ahu and leeward Hawai'i. Its wood is extremely hard. Hawaiians used it for fashioning *kapa* beaters, weapons, and *kahili* poles. Full sun is essential, as is a well-drained soil. It will provide good shade or can be used as a screen. It is wind and drought tolerant, but its tolerance of salt is unknown. (height 40 feet/canopy 30 feet)

Colvillea racemosa

COLVILLE'S GLORY **N**

ZONE 3

Fabaceae (Bean Family)

A superb tree from Madagascar, Colville's glory bears masses of bright orange spikes in October–November when most flowering trees have finished blooming. It will attain 50 feet in height. Plant it in any light, well-drained soil in full sun. It flowers best in hot coastal areas. It is moderately wind tolerant and partially drought tolerant and is a moderate grower. (height 50 feet/canopy 40 feet)

Schotia brachypetala

Conocarpus erectus

ZONE 3

BUTTONWOOD, SILVER BUTTONWOOD

Combretaceae (Combretum Family)

This very tough tree is native to the wide area from the Bahamas and southern Florida throughout the Caribbean to coastal Mexico and tropical West Africa. It is variously described as a low shrub to a tree reaching 50 feet in height. Local mature specimens of the species with dark green, leathery leaves and a highly attractive twisting, deeply furrowed trunk have reached 35 feet in height. Most commonly planted is a fuzzy, silver-leaved form from the Bahamas, which has attained 25 feet in height. We have opted to place this species in the medium tree category. The silver buttonwood retains the trunk and branching characteristics of the green-foliaged species. It is highly wind and salt tolerant and moderately drought tolerant. It forms an excellent windbreak or shade tree for beach gardens with strong onshore winds. Its foliage makes it a striking accent in the landscape, especially when night lighted. In the xeriscape, it has the same uses. Combine it with plants with dark green, shiny foliage or near a dark stone wall. It is a moderate grower. It must be grown from air layers to retain the silver color. (height 25 feet/canopy 50 feet)

Cordia subcordata

KOU

ZONE 3

Boraginaceae (Borage Family)

Native to coastal areas from East Africa to Polynesia, *kou* develops a dense, evergreen, rounded canopy providing good shade. It grows to 35 feet in height, producing light orange–colored flowers much of the year. They are used by the lei maker. It has moderate drought, wind, and salt tolerance. Use *kou* as a good shade tree, windbreak, or tall screen. It performs best in warm, leeward areas and is a moderate grower. (height 35 feet/canopy 50 feet)

Cupaniopsis anacardioides

CARROTWOOD, TUCKEROO

ZONE 3 EVALUATE

Sapindaceae (Soapberry Family)

Native to the broad geographical area from tropical Australia to Papua

New Guinea and eastern Indonesia, this tough tree will reach 35 feet in height. Although doing best in a well-drained soil, it will also perform satisfactorily in poor soils that are not well drained. It is drought, wind, and salt tolerant. It finds use as a shade tree, screen, or windbreak. It is a rather rapid grower. (height 35 feet/canopy 25 feet)

Cupressus sempervirens

ITALIAN CYPRESS

ZONE 3

Cupressaceae (Cypress Family)

Italian cypress is heat, drought, and wind tolerant, with some tolerance of mild salt air. It is native to the eastern Mediterranean and figures in classical mythology. Its stately, fastigiate form is used in the landscape as a strong vertical accent or may be used to form a tall hedge or screen. Although it may grow to a great height, it is of such slow growth that it is here included as a medium-sized tree. Its foliage may be used in wreaths or swags and is long lasting. It requires full sun and an open, well-drained soil. It is a rather slow grower. (height 50 feet/canopy 15 feet)

Delonix regia

ROYAL POINCIANA, FLAMBOYANT N

ZONE 3

Fabaceae (Bean Family)

var. *flava*

'Saipan Gold'

A wide-canopied tree from Madagascar reaching 50 feet in height, the royal poinciana is one of nature's most spectacular flowering trees. Flower color varies from a rich, deep red to medium red and brick red. There is a yellow-flowered variety, *D. regia* var. *flava*. The University of Hawai'i several years ago released another spectacular color form: 'Saipan Gold'. Flowers are a true gold (not yellow) and provide a brilliant splash of color in the dry garden. Blooming is extended over several months and peaks during warm summer periods. It performs best in hot coastal areas. Surface roots may extend as far as the canopy. Give the royal poinciana plenty of ground space in full sun. A moderate grower, it prefers a well-drained, light soil. It is quite wind, heat, salt, and drought tolerant. Use this tree as a strong color accent or for developing an eye-catching grove in large properties. Its almost ferny foliage provides light shade. (height 50 feet/canopy 100 feet)

Deplanchea tetraphylla

GOLDEN BOUQUET TREE

ZONE 2

Bignoniaceae (Catalpa Family)

Native to dry places in northeastern Australia and New Guinea, this evergreen species will reach 50 feet in height. It is a coastal tree, moderately tolerant of drought, and it will tolerate salt air but requires protection against salt spray and strong wind. It flowers all year, its bright yellow blossoms and large ornamental leaves providing a good accent and shade tree. Wreath makers value its ornamental dried seed pods. Plant it in full sun in an open, well-drained soil. It is slow growing. (height 50 feet/canopy 40 feet)

Elaeodendron orientale

FALSE OLIVE

ZONE 2

Celastraceae (Bittersweet Family)

Showing good wind tolerance and moderate tolerance to drought and salt, the false olive will rather slowly reach 40 feet in height, forming a dense canopy of oval-shaped, dark green leaves. Juvenile leaves are totally dissimilar, displaying slender leaves marked with mahogany red. It makes a good windbreak, screen, or general landscape tree for parks, schools, and medium-sized gardens. It can also be tubbed for interior use in a room with good light. The false olive tree is native to Mauritius. (height 40 feet/30 feet)

Erythrina abyssinica

RED HOT POKER TREE **T P N**

ZONE 3

Fabaceae (Bean Family)

Reaching 45 feet in height, this member of the bean family produces masses of scarlet flowers in late winter and spring. It is native to dry grassland savanna in Ethiopia, Kenya, Tanzania, and Uganda. It is moderately drought and wind tolerant and shows good resistance to the erythrina gall wasp. It is a valuable color accent in the xeriscape and a good shade tree. In its native areas, the wood is useful, and parts of the tree have medicinal value. Its furrowed bark is attractive. Plant this colorful tree where its bark may be seen and appreciated. It thrives in full sun in a well-drained soil. It is a fairly rapid grower. (height 45 feet/canopy 50 feet)

Erythrina dominguezii

BRAZIL CORAL BEAN **T P N**

ZONE 3

Fabaceae (Bean Family)

This is a handsome tree from deciduous dry forests of Bolivia, Peru, Paraguay, and Brazil. It grows to 50 feet and produces masses of salmon or coral pink flowers in late winter to early spring. It must have full sun and a well-drained soil. This is a startling color accent tree for large properties, parks, schools, and golf courses. Its canopy sheds a light shade, permitting the growth of other plants below. It is a moderate to slow grower and shows resistance to the erythrina gall wasp. (height 50 feet/canopy 50 feet)

Erythrina lysistemon

TRANSVAAL KAFFIRBOOM **T P N**

ZONE 3

Fabaceae (Bean Family)

A medium-sized tree reaching 40 feet in height, this South African species is native to dry savanna and sand dune areas. It bears many clear scarlet blossoms in winter and is a valuable shade and color accent tree. Plant it in any well-drained soil in full sun. It is heat, wind, and drought tolerant. Parts of the tree are employed medicinally in its native areas. It shows good resistance to the erythrina gall wasp. It is a moderate grower. (height 40 feet/canopy 40 feet)

Erythrina sandwicensis

WILIWILI, HAWAIIAN CORAL BEAN **T P N**

ZONE 4

Fabaceae (Bean Family)

This endemic Hawaiian deciduous tree is found in dry forest on all the major islands. It will reach 40 feet in height. It is highly drought and wind tolerant and moderately tolerant of salt wind. Its showy flowers appear when the tree is leafless and may be chartreuse, orange, yellow, white, and sometimes bicolored. Flowering time is variable, from summer in some areas to fall in others. The seeds also vary in color from brownish red to clear red. Both flowers and seeds are used in lei making. Its strong, light wood is still used for canoe outriggers and was formerly used for surfboards. Infestations of the erythrina gall wasp currently make the wiliwili a poor choice for the xeriscape, although its drought tolerance is excellent. A biocontrol, a wasp from Tanzania, East Africa (*Eurytoma erythrinae*), has been released by the Hawai'i State Department of Agriculture, with high expectations of success. Defer planting this fine endemic species until satisfactory control has been realized. As of this writing, great success has been shown in areas of the control wasp release. (height 40 feet/canopy 40 feet)

Erythrina vespertilio

BAT'S WING CORAL TREE **T P N**

ZONE 4

Fabaceae (Bean Family)

A tree from dry scrub areas of both eastern and northwestern Australia, this coral tree will reach 40 feet in height. Its interesting winged foliage is gray green and accented by bright orange red flowers that appear in winter. Its canopy is light, permitting the growth of other xerophytes beneath it. Its wood is light but tough. Surfboards were once made of it, and aborigines formerly used it to make shields, as it would not split if pierced by a spear. Its flowers, worn in the hair, are an aboriginal sign of mourning; roots are eaten raw. Unfortunately, it is attacked by the erythrina gall wasp (see notes about the wasp under *E. sandwicensis*). (height 40 feet/canopy 30 feet)

Eucalyptus bigalerita

NORTHERN SALMON GUM

ZONE 2

Myrtaceae (Eucalyptus Family)

New bark on this white-barked gum is a bright salmon pink, thereby giving rise to its common name. It is native to hot and seasonally dry places in the Northern Territory, Australia. It shows moderate tolerance of seasonal drought and is heat and wind tolerant. References vary as to its height, ranging from 20 to 50 feet, which is possibly due to differences in soil and water availability. Developing a rounded canopy of attractive foliage and showing masses of white flowers in winter and spring, this gum makes a good specimen tree, tall screen, or shade tree. It prefers a well-drained soil. It is a moderate grower. (height 50 feet/canopy 40 feet)

Euphorbia candelabrum

CANDELABRUM TREE **S T**

ZONE 4

Euphorbiaceae (Euphorbia Family)

Reaching slowly up to 50 feet in height, this species is highly drought, heat, wind, and salt air tolerant and is native to East Africa and Ethiopia. It is frequently found along Great Rift areas and in a range of elevations from sea level to 6,000 feet. Its name derives from its habit of branching, much like a giant candelabrum. In Ethiopia it has medicinal uses. Take care when handling this euphorb, as its thorns are sharp and its sap may cause skin irritation. Wash off sap as soon as possible with alcohol, as it is not water soluble. The candelabrum tree is valuable in both the xeriscape and coastal garden as a shade tree or large screen or windbreak. Plant it in full sun in a very well-drained soil and where people contact is unlikely. A similar but smaller species, the Malayan spurge tree (*Euphorbia antiquorum*), provides similar values in the xeriscape. Native to southern India, it grows to 25 feet in height, branches heavily, and produces a broad, dense canopy. It is also highly drought, heat, wind, and salt air tolerant and may be used as a small shade tree, hedge, screen, or windbreak. It is not figured. *E. candelabrum* (height 50 feet/canopy 50 feet); *E. antiquorum* (height 25 feet/canopy 25 feet)

Euphorbia haeleeleana

HĀʻELEʻELE SPURGE TREE

ZONE 3

Euphorbiaceae (Euphorbia Family)

An endemic species found in semidry places around Waimea Canyon, Kauaʻi, and in remnant dry forest in the Waiʻanae Range on Oʻahu, this is a handsome small tree that may slowly reach up to 40 feet in height. Leaves are somewhat succulent, producing a strong foliar accent in the xeriscape. Plant it in a well-drained soil in a place with full sun. It is heat and wind tolerant and moderately tolerant of drought. In addition to its shade, it finds use as a screen or windbreak. It is a slow grower. (height 40 feet/canopy 30 feet)

Ficus ingens

RED LEAF WILD FIG

ZONE 3

Moraceae (Mulberry Family)

Found in many parts of dry Africa from South Africa to sub-Saharan Africa, this fig will reach 40 feet in height. Its new foliage is reddish, making this species an interesting color accent. It is somewhat slow growing. Plant it in full sun in any well-drained soil. In the wild it also is found on rocky outcrops, even nearly vertical cliffs. Its fruit is small. It is heat, wind, and drought tolerant. Its broad canopy produces heavy shade and provides screening and windbreak. (40 feet/canopy 60 feet)

Ficus lyrata

FIDDLE LEAF FIG

ZONE 2

Moraceae (Mulberry Family)

Reaching 50 feet in height, this handsome tree from tropical West Africa thrives in either sun or shade in a well-drained soil. It has moderate salt, wind, and drought tolerance. It is prized for its large, decorative leaves. It is useful as a shade tree, screen, or windbreak. It is highly popular as a potted or tubbed specimen for indoor use. It is a moderate grower. (height 50 feet/canopy 50 feet)

Ficus palmeri

PALMER'S FIG

ZONE 4

Moraceae (Mulberry Family)

From very hot, dry places in western Mexico and Baja California, this 50-foot, wide-spreading tree produces a dense canopy and deep shade. Variously described as a small to large tree, it is reasonable to expect that in the xeriscape, given a well-drained soil and a modicum of water, it will eventually reach the greater size. It is a moderately slow grower. Use it as a shade tree, screen, windbreak, or accent tree in a large property. (height 50 feet/canopy 100 feet)

Ficus populifolia

POPLAR-LEAVED FIG

ZONE 4

Moraceae (Mulberry Family)

Ranging in the wild from dry parts of Central Africa to Kenya, Ethiopia, Yemen, and Saudi Arabia, this fig will reach 45 feet in height, forming a broad canopy. It is a moderate grower. It is wind, heat, and drought tolerant. Foliage is attractive on partially pendant branches. Plant it in full sun in any well-drained soil. Use it as a screening plant or foliage accent. It is a good shade tree. (45 feet/canopy 75 feet)

Ficus racemosa

CLUSTER FIG

ZONE 3

Morceae (Mulberry Family)

A tree from India, Southeast Asia, and tropical Australia, this fig may attain 45 feet in height, developing a broad canopy. The figs grow on the trunk and main branches. They are edible and attract birds. It is widely planted throughout India in varying environments but always in warm areas. It is highly prized in traditional Indian *ayurvedic* medicine, and its bark, roots, latex, and fruit are currently prescribed as treatment for a great number of physical problems. In Buddhism, the tree is referred to as the *udumbara* and is greatly revered. The cluster fig develops best in a well-drained soil planted in full sun. It is heat and wind tolerant and seasonally drought tolerant. It is a moderate grower and may be used as a fine shade tree, windbreak, or tall screen. (height 45 feet/canopy 60 feet)

Harpephyllum caffrum

WILD PLUM S

ZONE 3

Anacardiaceae (Mango Family)

Native to South Africa and north to Zimbabwe and southern Mozambique, this tough tree reaches

45 feet in height and withstands considerable drought and wind. Its fruits are edible and used to make jams and jellies and rose wine. The bark produces a dye and has medicinal characters. Its pale reddish wood is used in general construction and furniture making. Wild plum forms a dense canopy that is wind resistant; this species shows good drought tolerance and is a moderate grower. (height 45 feet/canopy 45 feet)

Jacaranda mimosifolia

JACARANDA **S**

ZONE 3

Bignoniaceae (Catalpa Family)

Reaching 50 feet in height, this colorful spring-flowering tree from northwestern Argentina shows good drought tolerance and moderate wind tolerance but poor salt tolerance. It is a moderately rapid grower and is deciduous. Its lilac blue flowers appear before new foliage develops. Although it will flower at low elevations, it performs far better in higher, cooler, dry elevations. Its seed pods are used in dry arrangements. Plant it in full sun in a well-drained soil. This is a good tree for large areas, golf courses, parks, and schools. It is a strong color accent. (height 50 feet/canopy 50 feet)

Kigelia africana

SAUSAGE TREE

ZONE 4

Bignoniaceae (Catalpa Family)

A unique species from tropical Africa, the sausage tree will grow moderately rapidly to 50 feet in height, forming an evergreen, rounded crown. Its reddish purple flowers open at night. It performs well in most soils with drainage in full sun. It shows good tolerance to drought, moderate to wind, and poor to salt. It makes an accent for the xeriscape, a conversation piece, and a good shade tree. The "sausages" do not fall from the tree but slowly wither away. The genus *Kigelia* is highly variable. Several species once listed as separate from *K. africana* have now been lumped into that one species. This includes an attractive variant thought to have come from Somalia (Somali sausage tree). Its horticultural requirements and uses are the same as above. It differs in having smaller, dark green foliage and smaller sausages. It is also a good choice for the xeriscape. (height 50 feet/canopy 50 feet)

Lagunaria patersonii

PRIMROSE TREE

ZONE 3

Malvaceae (Hibiscus Family)

This is an evergreen tree from the South Pacific and Australia, bearing gray green foliage and pink rose blossoms, hence the common name. It will grow to 40 feet in height in a rather columnar form. It is moderately drought resistant, salt tolerant, and wind tolerant. It fares well in any well-drained soil. This tree is a good choice for coastal areas with onshore winds but not on beachfronts. It makes a good color accent against a dark green background or use it as a tall hedge to screen off an unattractive view. It is a rather slow grower. (height 40/canopy 30 feet)

Lonchocarpus punctatus

DOTTED LANCEPOD, TRINIDAD GENO-GENO (P) (N)Z

ZONE 3

Fabaceae (Bean Family)

Native to dry parts of Trinidad and Tobago, Venezuela, and Colombia, this tree will grow moderately to 45 feet in height. It must have full sun and a well-drained soil. Flowers, borne in dense racemes, are lilac and appear in late summer or early fall. Use this colorful species as a shade tree or screen or color accent. It is heat and drought tolerant and moderately tolerant of wind and salt. A barbasco (fish poison) and an insecticide are derived from its roots. (height 45 feet/canopy 45 feet)

Lonchocarpus sericeus

GUAMA, GENO-GENO

ZONE 3

Fabaceae (Bean Family)

Reaching 45 feet in height, this deciduous West Indian and West African tree bears rose purple flowers in the spring. It shows good drought tolerance and moderate wind tolerance but no tolerance to salt. It requires good drainage and an open, friable soil. This is a good shade tree and color accent for parks and large properties such as schools. It is a moderate grower and may be briefly deciduous following a dry period. An insecticide and a barbasco are derived from its roots. (height 45 feet/canopy 45 feet)

Lonchocarpus violaceus

WEST INDIAN LILAC P N

ZONE 3

Fabaceae (Bean Family)

An evergreen, drought-tolerant species from arid areas of the West Indies and Colombia, this tree will reach 45 feet in height. Its flowers carry a fragrance similar to that of lilac and are borne in upright racemes in late summer or fall. They are rose violet, but white-flowering specimens are known. It shows moderate tolerance to salt and wind. Parts of this tree are poisonous and are used as a barbasco to stun and catch fish. Use it as a shade tree and as a bright color accent in the xeriscape. (height 45 feet/canopy 45 feet)

Lysiloma aurita

GUMARA PN

ZONE 4

Fabaceae (Bean Family)

Tropical dry forests of southeastern Mexico, Guatemala, and Nicaragua are home to this beautiful tree, which may reach 40 feet in height. Gracefully arching branches carry a thick, wide-spreading canopy of dark green, fine-textured foliage. Flowers, inconspicuous, are white and are seen in spring. Use gumara as a major shade tree in the xeriscape. It is drought, heat, and wind tolerant. Plant it in full sun in a well-drained soil. It is a moderate grower. Use it in the night garden with uplighting. (height 40 feet/canopy 60 feet)

Lysiloma latisiliquum

TAMARINDILLO, WILD TAMARIND N

ZONE 4

Fabaceae (Bean Family)

Native to the Bahamas, West Indies, and Mexico, this species will grow to 45 feet in height, developing a crown of fine-textured, almost lacy foliage. Flowers, seen in spring, are white, turning rose pink. It produces a pleasant, light shade permitting the growth of other species below. Its wood is hard and valued by the cabinetmaker. It is wind, heat, drought, and salt tolerant and is a moderate grower. It makes an excellent shade tree and can be used as a tall screening tree for parks and large gardens, for the xeriscape, and for the beach garden. Its fine-foliaged canopy is attractive with uplighting in the night garden. (height 45 feet/canopy 50 feet)

Manilkara zapota

CHICLE TREE

ZONE 3

Sapotaceae (Sapodilla Family)

Growing rather slowly to 40 feet in height, this valuable tree from southern Mexico and eastern parts of Central America has been under cultivation from pre-Colombian days and has spread throughout the Caribbean and tropical America. It is currently extremely popular in tropical Asia, where it is grown on a commercial scale. There are several fine cultivars available. Its foliage is a deep green in color, forming a dense canopy. Its fruit, called sapodilla, is prized for its delicate flavor, while its dark red wood is valued for its strength and durability and used in construction and for furniture and cabinetry. Tapping of the bark yields chicle, the white, sticky sap used in the production of chewing gum. The chicle tree is moderately wind tolerant and tolerant of heat, salt, and drought. It will grow in any well-drained soil, even dirty sand. In addition to its value in the edible garden, it can be used to form a hedge or screen or used as a solitary foliage accent. It is recommended for seaside gardens and the xeriscape. (height 40 feet/canopy 30 feet)

Melaleuca argentea

SILVER-LEAVED PAPERBARK

ZONE 3

Myrtaceae (Eucalyptus Family)

Attaining 50 feet in height, this showy tree displays a spreading canopy of silvery foliage, graceful pendant branches, and white papery bark. Strongly scented, creamy whitish flowers borne on cylindrical spikes are seen during the winter months. In its native northern Australia, its bark is used for wrapping food, as bedding, and as covers, while its leaves have medicinal uses as well as for flavoring meat. Plant this ornamental species in a well-drained soil in full sun. It is heat and wind tolerant and shows fair tolerance of drought. Use it as a specimen tree, windbreak, or tall screen or hedge. It is a moderate grower. (height 50 feet/canopy 40 feet)

Moringa drouhardii

DROUHARD MORINGA

ZONE 4

Moringaceae (Drumstick Tree Family)

Native to hot, very dry areas of southwestern Madagascar, this rapid-growing tree will reach 45 feet in height. Its white bark covering a bottle-shaped trunk is highly attractive, as are its abundant yellowish white flowers and clustered terminal plumes of fine, bright green foliage. It is highly drought tolerant and an excellent choice for the xeriscape. Use it as an accent specimen. For the night garden, it is dramatic when uplighting is provided. Plant it in full sun in an open, well-drained soil. Do not overwater. It will not tolerate wet feet. This tree has many other uses: an oil employed as a base for cosmetics and for medicinal purposes is expressed from its seeds; the strongly scented bark and wood are used in native medicine for the treatment of colds and coughs. (height 45 feet/canopy 25 feet)

Moringa stenopetala

NARROW-LEAVED MORINGA

ZONE 4

Moringaceae (Drumstick Tree Family)

Native to hot, dry places in the Rift Valley between Kenya and Ethiopia, this rare tree grows to a height of

40 feet, developing a broad canopy of fine-textured leaves that produce a useful light shade. Its pale yellow flowers are attractive, as is its whitish bark. It is an important food plant in southwestern Ethiopia and a source of oil and medicines. It is currently the subject of important research. Plant it in full sun in an open, well-drained soil. It is heat, wind, and drought tolerant and is a rather rapid grower. Use it in the xeriscape as an accent or shade tree. In the night garden with uplighting, this Moringa can be a major accent. It is a moderate grower. (height 40 feet/canopy 50 feet)

Noronhia emarginata

MADAGASCAR OLIVE **P**

ZONE 2 EVALUATE

Oleaceae (Olive Family)

Reaching 50 feet in height, this remarkable olive from Madagascar is highly heat, wind, and salt tolerant and moderately drought tolerant. It is widely used in beach gardens with an onshore, salt-laden wind, where it makes an excellent specimen, windbreak, hedge, or screen. It fruits heavily, which may present a maintenance problem. Plant it over a deep ground cover that produces heavy shade and may inhibit germination. The fruit pulp is edible, but the seed is considered toxic. It grows well in pure sand and good, well-drained soils and is at home inland as well as along the coast. It is a moderate grower. (height 50 feet/canopy 35 feet)

Olea europaea subsp. *cuspidata*

INDIAN OLIVE

ZONE 3

Oleaceae (Olive Family)

The wild relative of the common olive is native to Arabia and dry areas of southwestern Asia. Its plentiful fruit is not favored locally. Of the same size, 50 feet in height, the Indian olive finds good use as a windbreak, screen, or hedge and for its thick canopy, which provides good shade. It has the same horticultural requirements as the olive tree and is wind, heat, and drought tolerant, with a moderate tolerance to salt air but not salt spray. It is a moderate grower. (height 50 feet/canopy 50 feet)

Olea europaea subsp. *europaea*

OLIVE, ʻOLIWA

ZONE 3 EVALUATE

Oleaceae (Olive Family)

Native to the Mediterranean region, the olive tree will grow slowly to 50 feet in height, developing a round-headed, evergreen canopy of soft, gray green foliage. It tolerates drought, heat, wind, and poor soils. It does not bear fruit at low elevations in Hawaiʻi, which eliminates a maintenance problem. It does have a tendency to produce suckers, which must be kept cut away. It has been in cultivation for more than four thousand years and is noted in some of man's earliest writings. Greek mythology relates that Athens was named for the goddess Athena, who produced the olive, the gift most useful to man. It is the Greek symbol for victory. Olympic winners were crowned with a wreath of olive leaves. Its foliage is the Christian symbol of peace. It was an olive branch carried by a dove to Noah signaling God's forgiveness and the end of the great flood. The olive provides good shade, windbreak, and screening, and is a brilliant accent when uplighted in the nightscape. (height 50 feet/canopy 50 feet)

Peltophorum pterocarpum

YELLOW POINCIANA

ZONE 2

Fabaceae (Bean Family)

Native to Southeast Asia and Australia, this rapid-growing evergreen species will reach 50 feet in height, developing a broad canopy of fernlike foliage. Flowering profusely in late spring to midsummer, it produces a high color accent. Its fragrant yellow flowers are followed by colorful, persistent, flat reddish seed pods that eventually turn black. Unfortunately, this colorful tree is subject to wind damage and should be planted away from buildings or pedestrian areas. It shows only moderate wind and salt tolerance but is tolerant of heat and moderately tolerant of drought. In its native areas, the timber is used for cabinetry, medicinally, and as a source for a brown dye used to color batik. It is well used as a colorful accent or as a grove bordering a large property. (height 50 feet/canopy 50 feet)

Pithecellobium dulce 'Variegata'

WHITE ʻOPIUMA T N

ZONE 4 H(HPWRA)

Fabaceae (Bean Family)

Growing fairly rapidly to 40 feet in height, this tropical American tree is valued for its variegated foliage, which, when new, is pink, then turns green and white. It grows readily in any well-drained soil and is drought and wind tolerant. Prune with care to avoid its sharp thorns. It makes a good shade tree, screen, or windbreak and is beautiful in the night garden. (height 40 feet/canopy 40 feet)

Pimenta dioica

ALLSPICE TREE

ZONE 3 H(HPWRA)

Myrtaceae (Eucalyptus Family)

Possibly native to the area ranging from southern Mexico to Central America, Cuba, and Jamaica, the allspice tree reaches 50 feet in height. It is prized for its dried unripe fruit, which is the source of the allspice of commerce due to the fragrance of its fruit, which is reminiscent of cloves, cinnamon, and nutmeg. Its leaves are also fragrant. Plant it where the foliage is within reach, as people like to pluck a leaf and savor the scent. A moderate grower, the tree thrives in a well-drained soil and will bear large amounts of fruit. It is best to grow it over a fairly deep ground cover that will allow the fruit to fall to the ground where no pedestrian traffic is expected. Allspice is heat and wind tolerant and moderately tolerant of drought but not salt. Its ornamental bark is particularly attractive in the night garden with uplighting. (height 50 feet/canopy 35 feet)

Pimenta racemosa

BAY RUM TREE

ZONE 3 EVALUATE

Myrtaceae (Eucalyptus Family)

Attaining 40 feet in height, the bay rum tree is native to dry slopes of Puerto Rico, Cuba, and smaller Caribbean islands to Venezuela and the Guiana region. Its bark is colorful and the dark green foliage is fragrant. It should be planted where both qualities can be enjoyed. It is commercially grown as a source of myrcia oil (bay oil), from which bay rum is derived and used in cosmetics and medicines. It is partially drought tolerant and wind tolerant. Use it as a screen, tall hedge, or colorful accent in the landscape. In the night garden, uplighting will make the colorful bark glow. A well-drained soil in full sun will produce best results. (height 40 feet/canopy 30 feet)

Piscidia piscipula

ZONE 3

JAMAICAN DOGWOOD, FISH POISON TREE

Fabaceae (Bean Family)

A rather slow grower reaching 45 feet in height, this handsome tree is native to both wet and dry parts of southern Florida and dry coastal regions of the West Indies, Mexico, and Central America. It bears a large crop of light pink flowers followed by interesting papery seed pods. It is salt wind tolerant and moderately tolerant of drought. Its wood is hard and heavy. It is used in construction. The name "fish poison tree" is derived from its use as a barbasco for catching fish. When its bark is thrown into a still pond, a chemical is released that temporarily narcotizes fish. They float to the surface to be easily caught. It does not harm the fish nor those who eat them. (height 45 feet/canopy 30 feet)

Plumeria obtusa

SINGAPORE PLUMERIA P

ZONE 3

Apocynaceae (Dogbane Family)

Its common name is misleading, for the Singapore plumeria is native to the West Indies. It will reach 35 feet in height, bearing large clusters of white, fragrant flowers from spring through fall. The glossy, dark green foliage is partially evergreen. It will grow in almost any well-drained soil in full sun. It is moderately drought tolerant. Use this species as a flowering accent where its foliage is set off against a light background or structure. Flowers are used in arrangements but are too large and floppy for lei making. It withstands onshore winds if given minimal protection. It is a moderate grower. (See also Bahamas plumeria in "Small Trees.") (height 35 feet/canopy 40 feet)

Plumeria rubra

PLUMERIA, TEMPLE TREE **P**

ZONE 3

Apocynaceae (Dogbane Family)

'Tillie Hughes'

'Keiki'

'Hilo Beauty'

'Pinwheel'

'Bali Whirl'

'Princess Victoria'

Prized for its fragrant flowers used to make leis, this tropical American species will reach 35 feet in height. Well over a hundred flower color variations ranging from white to light and dark pink, orange, yellow, and gold, to red and dark red bicolors, semidoubles, and dwarf forms are available. Trees drop their foliage in late fall and produce great clusters of flowers just before refoliating in early spring, continuing flowering until late fall. Plumeria grows well in any well-drained soil in a sunny location. It is only moderately drought, wind, and salt tolerant. Plumerias are valued in the garden for shade and as color accents. It is a moderate grower. Entire branches may be cut for decorating and continue flowering. In Aztec times, the flowers were used in ceremonials, and once consecrated, death was the penalty for a commoner who touched them. Virgins of noble Aztec families wore plumeria flowers in their hair. This reverence has been carried into modern times, as garlands of plumeria flowers are used in the month of May in Taxco, Mexico, to present to the Virgin Mary. The plumeria is Hawai'i's most popular homegrown lei flower. (height 35 feet/canopy 40 feet)

Podocarpus elatus

BROWN PINE, PINE-TINT PODOCARPUS

ZONE 3

Podocarpaceae (Podocarpus Family)

Due to its slow growth rate, this large tree is here listed as a medium tree. Native to Australia, it shows moderate drought and salt tolerance and good wind tolerance once well established. It accommodates to any well-drained soil in both sun and light shade. Use it as a shade tree, hedge, or screen or a dark background for light-foliaged specimens. (height 50 feet/canopy 30 feet)

Podocarpus macrophyllus

JAPANESE YEW, KUMASAKI
SOUTHERN YEW

ZONE 2

Podocarpaceae (Podocarpus Family)

Slowly reaching 50 feet in height with a broadly columnar form, this Japanese evergreen species thrives in full sun or shade in a well-drained soil and, although preferring moisture, is moderately tolerant of drought and salt once well established. It has good wind tolerance. Japanese yew makes an excellent hedge or screen or a vertical accent. (height 50 feet/canopy 30 feet)

Prosopis pallida

THORNLESS KIAWE, ALGAROBA N

ZONE 4

Fabaceae (Bean Family)

Thriving in hot, dry areas, this tree from arid areas of northern Peru will reach 50 feet in height. In full sun and in well-drained soil, *kiawe* produces a fine, light shade permitting the growth of other xerophytes under its canopy. It is a nitrogen fixer. Its attractive bark is furrowed, twisted, and rough. Its wood is heavy, dense, and produces excellent charcoal. Most *kiawe* trees bear sharp spines. Fortunately for the xeriscaper, there is a spineless form that is readily propagated by cuttings or air layering. It is heat, drought, and moderately salt and wind tolerant and a rather rapid grower. (height 50 feet/canopy 50 feet)

Pseudobombax ellipticum

PINK BOMBAX, SHAVING-BRUSH TREE

ZONE 3

Bombacaceae (Bombax Family)

Unusual flowers, looking much like an old-fashioned shaving brush, appear on this tropical Mexican tree during its leafless period from late winter to early spring. New leaves are mahogany red, turning green while maturing. In addition to its pink flowers, it has a very ornamental, green, fissured bark. It grows best in full sun in a fertile, well-drained soil. It has moderate drought, wind, and salt tolerance. There is a white-flowered form and one with dark pink blossoms. Use this plant as a wonderful seasonal color accent where both its flowers and bark may be enjoyed. It may reach 40 to 50 feet in height. It has traditional medicinal uses: a tea of the flowers is used to relieve fevers and coughs; the bark and roots are used to treat toothache. It is a slow grower. (height 50 feet/canopy 50 feet)

Sapindus oahuensis

LONOMEA, AULU

ZONE 3

Sapindaceae (Soapberry Family)

Endemic to Kaua'i and O'ahu, this species is found in mesic to dry forest at fairly low elevations. It also grows well at sea level under cultivation. It will reach 45 feet in height, forming an attractive shade tree that shows moderate tolerance to drought. Its seeds are made into leis. See remarks under *Sapindus saponaria* relative to toxicity. Give this species full sun and a well-drained soil. It produces a good shade tree for the dry garden and is a moderate grower. (height 45 feet/canopy 35 feet)

Sapindus saponaria

SOAPBERRY, A'E

ZONE 3

Sapindaceae (Soapberry Family)

Native to the wide-ranging area from tropical Mexico to Hawai'i, to the South Pacific including New Caledonia, and to Africa, *a'e* is locally found in high, mesic forest but will also thrive under cultivation at sea level. In full sun in a well-drained soil it will reach 50 feet in height, showing moderate drought tolerance and good wind tolerance. It makes a fine shade tree with a light crown. It is briefly deciduous. Its jet black seeds are made into leis. Care must be taken by the lei maker, as the seeds contain a toxin. A protective face mask is advisable when drilling the seeds. It is a moderate grower. (height 50 feet/canopy 50 feet)

Schotia brachypetala

FUCHSIA TREE **N**

ZONE 3

Fabaceae (Bean Family)

Cassia sieberiana

Native to bushveld areas of south central Africa and Mozambique, this highly ornamental species grows to 40 feet in height, developing a colorful, very dense, rounded canopy. New foliage is rose colored, turning ruby red, then copper, and finally a rich dark green. Flowers occur in clusters and are tubular, deep red, with long protruding stamens. It is a slow grower. Use it as a good screen, windbreak, shade tree, or color accent. It is moderately drought, heat, and wind tolerant. (height 40 feet/canopy 40 feet)

Sterculia africana

AFRICAN STAR CHESTNUT

ZONE 3

Sterculiaceae (Cacao Family)

An attractive tree to 50 feet in height native to dry acacia bushland in southern and eastern Africa, this species develops a many-branched, shade-giving canopy. Its mottled bark is handsome and provides a fiber used to make string. It must have full sun and a well-drained soil. This species is heat and wind tolerant and moderately tolerant of drought. It is partially deciduous. Its greenish yellow flowers are tinged and veined in red and are seen during the brief leafless period in spring. Use it for its shade, as a windbreak, or as a tall screen. It is a moderate grower. Its bark is handsome when lighted at night. (height 50 feet/canopy 50 feet)

Terminalia catappa

TROPICAL ALMOND, FALSE KAMANI, KAMANI HAOLE

ZONE 3

Combretaceae (Combretum Family)

Usually associated with seaside sites, the tropical almond also grows inland. It is native to the East Indies and is a wide-spreading tree growing to 50 feet in height. It is briefly deciduous in winter, its leaves turning gold or red prior to dropping. It has excellent tolerance of wind and salt and is moderately tolerant of drought. This is a great shade tree for the large property or parks, especially in coastal sites with onshore wind. (height 50 feet/canopy 100 feet)

Thespesia populnea

MILO, PORTIA TREE
ZONE 3 H(HPWRA)
Malvaceae (Hibiscus Family)

Probably native to the Old World tropics, this tree reaches 40 feet in height and is associated with warm, leeward coastal gardens. It is probably an introduction from the south during the migration period. It shows good heat and wind tolerance and moderate drought tolerance. Its large crop of bright yellow flowers may be made into a beautiful lei, while its wood and fiber have traditional Hawaiian uses. Use it as a good, rather rapid-growing shade tree or windbreak. Plant it in full sun in a well-drained soil. (height 40 feet/canopy 50 feet)

LARGE TREES OVER 50 FEET

Acacia xanthophloea

FEVER TREE N

ZONE 3

Fabaceae (Bean Family)

This magnificent tree, native to the tropics of East Africa and south into temperate areas of Zambia, Zimbabwe, Swaziland, and Botswana, will reach almost 100 feet in height, developing a broad canopy. It is outstanding for the striking light yellow green color of its bark and for the fact that the bark is capable of photosynthesizing. New growth bears conspicuous 2- to 3-inch white, double spines that are readily avoided. An abundant crop of sweet-scented, yellow, ball-shaped flowers appears in spring and summer. Its fine foliage provides food for elephants and giraffes, an important fact for the xeriscaper. Leaflets fold up at night and during periods of intense heat, an ability shared by many species. The tree casts a light shade. Although recommended for the xeriscape in protected areas of the U.S. southwest where temperatures do not fall below 25 degrees Fahrenheit, the fever tree in nature is usually associated with warm areas with a high water table. Under cultivation it shows moderate tolerance to drought but thrives with minimal watering during dry months. It can, however, successfully withstand periods of drought and is an excellent xeriscape choice for large properties. It grows moderately rapidly and fares best in a deep, light, well-drained soil. Plant the fever tree in full sun. Uplight the trunk for the nightscape to produce a striking, glowing accent. (height 100 feet/canopy 100 feet)

Adansonia digitata

BAOBAB TREE

ZONE 4

Bombaceae (Bombax Family)

Native to many dry parts of Africa from sub-Saharan areas to East Africa and south to northern South Africa, the baobab tree will reach 90 feet in height. Its base becomes greatly swollen with water during the brief rainy season and acts as a reservoir during the dry season. It will grow in any well-drained soil and tolerates heat, wind, and drought. It is a good choice for large properties, parks, and golf courses, forming a highly attractive accent. Tender leaves may be eaten and contain calcium and iron, while the fruit provides a source of vitamins B_1 and C. Strong, durable fiber is extracted from the bark. In some parts of Africa the tree has religious significance. It is a rather slow grower. (height 90 feet/canopy 70 feet)

Andira inermis

PARTRIDGE WOOD, CABBAGE BARK **P N**

ZONE 3

Fabaceae (Bean Family)

Reaching 100 feet in height, this tree from tropical America and tropical West Africa produces a valuable timber. Both its seed and bark are toxic. Its clustered fragrant flowers are deep lilac to deep rose pink in color and appear intermittently throughout the year. It shows moderate drought, wind, and salt tolerance. It can be used as a colorful shade tree, windbreak, or screen. It is not particular as to soil as long as it is well drained. It is a moderate grower. (height 100 feet/canopy 75 feet)

Araucaria columnaris

COOK PINE

ZONE 3

Araucariaceae (Araucaria Family)

Native to the Isle of Pines near New Caledonia, this fast-growing evergreen tree may reach well over 100 feet in height. As its name implies, it has a columnar form. It is wind tolerant and is moderately tolerant of salt and drought, as is its closely related cousin, Norfolk pine (*A. heterophylla*) (not pictured). Their columnar form makes them excellent accents against large buildings and as windbreak, as their canopy does not require a large space. These are for large landscapes. (height 100 feet/canopy 30 feet)

Azadirachta indica

NEEM TREE

ZONE H(HPWRA)

Meliaceae (Mahogany Family)

Native to India and Southeast Asia, the highly esteemed neem tree will grow to 60 feet in height. It will prosper in any well-drained soil and tolerates heat, wind, and drought and a small amount of salt. It is the source of an antiseptic oil and has countless medicinal applications. Its highly ornamental leaves have many insecticidal properties and form a dense crown, making it a fine shade tree, screening tree, or windbreak. It is a moderate grower. (height 60 feet/canopy 60 feet)

Barnebydendron riedelii

CARDINAL TREE

ZONE 3

Fabaceae (Bean Family)

Native to hot, semidry places from Guatemala to Honduras, this partially deciduous species will grow rapidly to 100 feet in height. During its brief, almost leafless period in winter and early spring, it becomes covered with masses of bright scarlet flowers. It fares best in dry leeward areas. It shows moderate drought, heat, wind, and salt tolerance. Use this tree as a powerful color accent in the large landscape, in parks, and around government buildings, schools, and golf courses. It is also a good screening tree and windbreak. (height 100 feet/canopy 100 feet)

Bombax ceiba

RED SILK COTTON TREE

ZONE 3

Bombacaceae (Bombax Family)

A large deciduous tree from tropical Asia and east to New Guinea reaching 80 feet in height, this species is grown for its spectacular spring flowering. Its fleshy flowers are a strong red, but seedlings with bright orange coral blossoms are not uncommon (yellow seedlings also rarely occur). It is a fairly rapid grower in full sun in a good, well-drained soil. The flowers, which may be cooked as a vegetable, are followed by large pods filled with silky hairs called "silk cotton" and numerous small, black seeds. The silk cotton is used to stuff pillows and mattresses. Young growth may display small, sharp spines that are absorbed as trunk and branches grow. It is wind and heat tolerant and moderately drought tolerant but shows little tolerance to salt. It may be used singly as a highly colorful accent or in a park, golf course, resort, or large school campus or planted as a grove. (height 80 feet/canopy 80 feet)

Brosimum alicastrum

MAYAN BREAD NUT

ZONE 3

Moraceae (Mulberry Family)

Native to dry forests from Mexico to Central America, the West Indies, and northern South America, this rapid-growing evergreen tree will reach up to 100 feet in height. It forms a dense, dark green canopy that produces an edible seed high in starch, proteins, and vitamins A and C. They may be eaten boiled, with a taste like chestnuts, or roasted and ground as a coffee substitute. Fruits ripen usually in late spring or summer. They formed a significant component of the ancient Maya diet, and the tree was cultivated by the Maya for that purpose. Its wood is useful. Plant Mayan bread nut in any open, well-drained soil in full sun. It is wind, heat, and drought tolerant and moderately tolerant of salt. Use this tree as a large windbreak, screen, boundary planting, or for its edible fruit. Plant it where its fruit will not fall on paving. (height 100 feet/canopy 75 feet)

Bucida buceras

GEOMETRY TREE S

ZONE 3

Combretaceae (Combretum Family)

Reaching 60 feet in height, the geometry tree will tolerate both coastal and inland environments, planted in a well-drained soil or even sand. It is moderately salt, wind, and drought tolerant. Spines on immature specimens may decrease and disappear as the tree matures. It is a good shade tree, screen, or tall hedge and windbreak. It is native to Florida, the West Indies, and Central America. (height 60 feet/canopy 60 feet)

Bursera simaruba

GUMBO LIMBO, WEST INDIAN BIRCH

ZONE 3

Burseraceae (Torchwood Family)

Native to the wide area from southern Florida to the West Indies, Central America, and northern South America, this rapid-growing, unusual tree will reach 75 feet in height. Its bark is reddish and peels away, revealing a light gray underbark. It is drought, salt, and wind tolerant. It requires a well-drained soil and will not tolerate soggy soils. It makes a fine color accent tree and has been used as a boulevard tree where planting spaces are broad. A resin from the trunk is used to make glue, varnish, a waterproofing substance, and incense. It is related to frankincense and myrrh. It is a moderate grower. (height 75 feet/canopy 90 feet)

Byrsonima crassifolia

NANCE

ZONE 3

Malpighiaceae (Malpighia Family)

Native from dry places in Mexico, Central America, Caribbean islands, and south to Brazil and Paraguay, nance is wind and drought resistant. It bears showy yellow flowers that turn red on withering. It is prized for its yellow fruit, which may be eaten raw or cooked to make a dessert. In Costa Rica, a rumlike liquor called Crème de Nance is produced from the fruit, which also contains a useful, edible fat. Use nance in the xeriscape as a color accent, high screen, or windbreak. Plant it in a well-drained soil in full sun. It will attain 60 feet in height. (height 60 feet/canopy 60 feet)

Calophyllum inophyllum

KAMANI, ALEXANDRIAN LAUREL P

ZONE 2 EVALUATE

Clusiaceae (Mangosteen Family)

Reaching 60 feet in height with a broad canopy, this slow-growing evergreen tree is native to coastal areas of the Indian and western Pacific oceans. Clusters of fragrant white flowers appear in spring and summer and are used to make leis. Its leathery leaves and spherical, green fruit are used in both green and dry arrangements, swags, and wreaths. It is an excellent shade tree, windbreak, or screen. It is tolerant of salt, most soils, wind, and is moderately tolerant of drought. (height 60 feet/canopy 80 feet)

Cassia x *nealiae*

RAINBOW SHOWER N

ZONE 3

Fabaceae (Bean Family)

'Wilhelmina Tenney'

'Queen's Hospital White'

'Lunalilo Yellow'

'Nii Gold'

Growing rapidly to 60 feet in height, this is one of the landscaper's favorite trees. It is a Hawai'i-bred cross between *Cassia fistula* and *Cassia javanica* and flowers abundantly from spring to fall. There are four named selections, all with persistent flowers: 'Wilhelmina Tenney', with bicolored blooms of deep cerise and yellow, is the most frequently planted cross and is the official tree of the City and County of Honolulu. It has not been seen to produce seed pods. 'Lunalilo Yellow' has flowers opening yellow orange, fading to a bright yellow with age. A few seed pods may appear but are not a problem. 'Queen's Hospital White' displays flowers opening a pale yellow, fading to almost white. A few seed pods may follow. 'Nii Gold' is a sport of 'Wilhelmina Tenney', with deep gold flowers, fading to strong yellow. Unlike the other rainbow showers, 'Nii Gold' has only a monthlong flowering period, which may occur anytime from spring to fall. It produces very few seed pods. Rainbow showers fortunately inherit the drought tolerance displayed by one parent, *C. fistula*. They are also wind tolerant but not strongly salt tolerant, although they will perform well just a block in from the ocean. They are not particular about soil and will thrive even in sand. Rainbow showers are spectacular color accents and provide good shade. (height 60 feet/canopy 60 feet)

Cochlospermum vitifolium

BUTTERCUP TREE

ZONE 3

Bixaceae (Arnotto Family)

'Plena'

Native to dry areas from northern Mexico south to Venezuela, Peru, Bolivia, and to northern Brazil, this winter-flowering tree bears large, single yellow flowers during its brief deciduous period. There is also a double-flowered form. It is drought tolerant but not salt tolerant. It must have good drainage. Planted in full sun, it will eventually reach 60 feet or more in height. This is a good color accent in the xeriscape when other colorful trees are not in flower. It is a rather slow grower. (height 60 feet/canopy 50 feet)

Dalbergia latifolia

INDIAN ROSEWOOD, BLACKWOOD N

ZONE 3 EVALUATE

Fabaceae (Bean Family)

A large deciduous tree from both dry and mesic areas of India, this species may reach 100 feet in height when growing in a rich, deep soil with ample rainfall. In dry areas of India, it reaches 70 feet in height, with abundant, pendant inflorescences of showy, pea-shaped white flowers. It is tolerant of moderate drought, heat, and wind. Its richly colored, fragrant heartwood is prized for making furniture and paneling and is reported to be 2.5 times as hard as oak. Its bark contains tannin and has medicinal properties. Its fine foliage is highly attractive, providing a light shade. A good choice for large properties, parks, and golf courses, it is a nitrogen fixer and grows satisfactorily in any well-drained soil. Its foliage is high in nitrogen and is used for composting. Use this beautiful tree as an accent, specimen, windbreak, or tall screen. Its canopy is sufficiently open to permit good understory planting. It is a rather rapid grower. (height 100 feet/canopy 75 feet)

Enterolobium cyclocarpum

EARPOD, ELEPHANT'S EAR N

ZONE 3

Fabaceae (Bean Family)

A fast-growing species from tropical America, this tree will reach 100 feet in height with an even greater spread.

It is briefly deciduous. Plant it in full sun in any soil with good drainage. It is a nitrogen fixer. It has good drought tolerance but only moderate tolerance of wind and salt. It is an excellent shade tree if the landscape will accommodate its size. Seed pods and seeds are used in arrangements and the latter for making jewelry. (height 100 feet/canopy 150 feet)

Erythrina addisoniae

ADDISON CORAL BEAN **T P N**

ZONE 3

Fabaceae (Bean Family)

A large tree reaching 60 feet in height, this coral bean is native to dry areas of northern Ghana and neighboring countries. It produces abundant bright red flowers in summer when planted in full sun and in a well-drained, open soil. It is a moderate grower. Plant it for its shade-giving quality, as a strong color accent, or as a screen or windbreak. It is resistant to the erythrina gall wasp. Its bark is being researched for suspected medicinal properties. It is tolerant of wind and heat and moderately drought tolerant. (height 60 feet/canopy 60 feet)

Erythrina sacleuxii

KENYA CORAL BEAN **T P N**

ZONE 3

Fabaceae (Bean Family)

A colorful species from dry areas of Kenya and Tanzania, this is a large tree, reaching 60 feet in height. During winter months, it produces quantities of bright red and white flowers. It is deciduous but produces a dense, rounded crown most of the year. It requires full sun in a well-drained soil. Use it as a color accent for the winter garden or as a screen or windbreak. It is wind, heat, and moderately drought tolerant. It is resistant to the erythrina gall wasp. It is a rather rapid grower. (height 60 feet/canopy 60 feet)

Erythrina variegata var. *orientalis*

INDIAN CORAL TREE, TIGER'S CLAW, WILIWILI HAOLE **P T N**

ZONE 3

Fabaceae (Bean Family)

'Alba'

Erythrina variegata

Erythrina variegata

This species is native to the broad area from India to southern and eastern Polynesia. Growing moderately rapidly to 60 feet in height, it bears masses of brilliant red flowers in winter while briefly leafless. There is a white-flowered form. *Erythrina variegata,* with green and yellow variegated leaves, is similar in all respects. Both are nitrogen fixers and will tolerate almost any well-drained soil. Both have good wind tolerance and good tolerance to salt and drought. Use them to create a shady, colorful grove or as a color accent. Unfortunately, this species is subject to severe damage by the erythrina gall wasp, and its use should be deferred until controls have been established (see note about the erythrina gall wasp in "Small Trees"). (height 60 feet/canopy 70 feet)

Eucalyptus camaldulensis

RIVER RED GUM

ZONE 2

Myrtaceae (Eucalyptus Family)

Although in its native Australia this gum is associated with watercourses, it is also listed as moderately drought resistant. It will grow rapidly to 135 feet in good, well-drained soil. It is a graceful species with pendant branches and leaves. Its wood is used for flooring and firewood. Parts of the tree have medicinal uses. River red gum is a good selection for large properties and parks, providing light shade and screening. (height 135 feet/canopy 75 feet)

Eucalyptus citriodora

LEMON-SCENTED GUM

ZONE 2

Myrtaceae (Eucalyptus Family)

A tall, slender tree from Australia, this rapid-growing evergreen species will reach 75 feet in height, developing an open crown that provides light shade. Its foliage is strongly lemon scented. White flowers appear in winter. Give it space in a well-drained soil to produce an excellent shade tree or to develop a grove. Plant it where wind will bruise its leaves, releasing its pleasant aroma. It has moderate drought and wind tolerance. It has no salt tolerance. (height 75 feet/canopy 50 feet)

Ficus benghalensis

INDIAN BANYAN, BANYAN, VADA TREE

ZONE 3

Moraceae (Mulberry Family)

This is a huge species, growing to 80 feet in height with an extremely broad canopy. It is capable of covering several acres, spreading by its aerial roots that take root and develop into additional trunks. Clearly, this is a singular tree for very large properties and can be planted as a signature tree, an unusual focal point in the landscape, or as a large windbreak or screen. Native to India and Pakistan, it grows in almost any well-drained soil and has moderate drought and salt tolerance and good wind tolerance. It is sacred to Hindus. (height 80 feet/canopy several acres)

Ficus benjamina

BENJAMIN TREE, BENJAMIN FIG, WEEPING FIG

ZONE 2 EVALUATE

Moraceae (Mulberry Family)

var. *comosa*

'Starlight'

Native from India and Malaysia, this fast-growing tree reaches 60 feet in height, forming a broad, dense, weeping canopy twice that measurement. Foliage is variable, ranging from very slender to moderately slender leaves. It makes an excellent large shade tree, focal point, or windbreak. It tolerates most soils as long as it has good drainage, and it has moderate drought, salt, and wind tolerance. A variant, *Ficus benjamina* var. *comosa,* bears yellow fruit rather than red. *F. benjamina* 'Starlight' is one of several green and white leaf varieties. It is a rapid grower and requires ground space, as it produces wide-spreading surface roots. It is a fine selection for parks, golf courses, and other large landscapes. (height 60 feet/canopy 120 feet)

Ficus macrophylla

MORETON BAY FIG

ZONE 3 EVALUATE

Moraceae (Mulberry Family)

This eastern Australian Fig will reach 60 feet in height with an equal spread. It is fast growing, succeeding in almost any well-drained soil and is tolerant of smog, vog, wind, drought, and salt. Its dense canopy makes it ideal for use as a screen or windbreak and as mitigation against noise. It is a good choice for parks and golf courses in hot, dry leeward areas. In addition to canopy space as noted above, it must have ground space, as it develops buttresses and wide-spreading surface roots. (height 60 feet/canopy 60 feet)

Ficus religiosa

BO TREE, PEEPUL TREE

ZONE 3

Moraceae (Mulberry Family)

Native to India and cultivated throughout the country for its religious importance to both Hindus and Buddhists, this large fig will reach 100 feet in height, producing a wide-spreading canopy. It has moderate drought tolerance and good heat and wind tolerance. This is a magnificent accent tree for parks, golf courses, and large estates. It needs full sun and a well-drained soil. It does not develop aerial roots or surface roots. It is a moderate grower. (height 100 feet/canopy 100 feet)

Ficus sycomorus subsp. *gnaphalocarpa*

LARGE FRUITED SYCAMORE FIG

ZONE 4

Moraceae (Mulberry Family)

Native to widespread arid areas of Africa, this fig may reach 120 feet in height, with a broad canopy. A tough species, it is drought, heat, and wind tolerant and will perform well in any well-drained soil. It is a good choice for parks and golf courses, providing shade, windbreak, or screening in large, dry leeward areas. Its fruit is edible. It is closely related to the sycamore tree of the Bible, and although light, its wood has been found in mummy cases in ancient Egypt. There is considerable taxonomic disagreement involving this tree. One reference considers *F. gnaphalocarpa* a synonym of *F. sycomorus*, while other taxonomists list it as a subspecies of *F. sycomorus*. In any event, it is a fine selection for a large xeriscape. Its light-colored bark is striking with lighting at night. It is a moderate grower. (height 120 feet/canopy 120 feet)

Ficus vasta

SOCOTRA FIG

ZONE 4

Moraceae (Mulberry Family)

A large fig from dry areas of East Africa, Somalia, Yemen, and the island of Socotra, this tree will grow to 75 feet in height. It is drought, heat, and wind tolerant. It is a good choice for the large xeriscape, parks, or golf courses. Give it plenty of room; both its canopy and root system become wide spreading. It may become deciduous during long dry spells. It makes an excellent cover for a rocky cut or hillside to prevent erosion, to screen an unfortunate view, or to provide noise blockage. Performing best in a well-drained soil, the Socotra fig will also thrive in virtually any soil or rocky area with good drainage and in full sun. Its fruit is edible. (height 75 feet/canopy 150 feet)

Hildebrandia barteri

KARIYA

ZONE 2

Sterculiaceae (Cacao Family)

Native to tropical Africa from the Ivory Coast to Cameroon, this species grows to 100 feet in height. It is found on the drier type of lowland rainforest and is valuable for use in large xeriscapes in the zone requiring a modicum of irrigation. It is a moderate grower. Flowers, produced abundantly in winter when the tree is temporarily leafless, are red and may be seen from a distance. Plant this large tree in full sun in an open, well-drained soil. It is a fine color accent, shade tree, or screen. It is a moderate grower. (height 100 feet/canopy 75 feet)

Lonchocarpus domingensis

SAVANETTE **N**

ZONE 3

Fabaceae (Bean Family)

From dry areas of Mexico, Dominican Republic, Jamaica, Granada, and Trinidad, this tall, slender tree will reach 60 feet in height. Its flowers are light purple, pinkish, or white, appearing in spring and summer. It produces a useful timber. Give it full sun in a well-drained soil. Use it as a shade tree, tall screen, or windbreak. It is moderately drought, heat, and wind tolerant and a fairly rapid grower. (height 60 feet/canopy 40 feet)

Mangifera indica

MANGO **S**

ZONE 3

Anacardiaceae (Cashew Family)

Native to tropical Asia, the mango tree will attain 120 feet in height. It fruits best in a well-drained soil in lowland, hot, dry areas. Care should be taken in handling any part of the tree, as some people are allergic to its sap. There are literally hundreds of varieties of mango. Semidwarf varieties have been developed, and much work is currently being done in India to develop even more. There are two locally available semidwarf varieties. 'Julie' slowly grows to 20 feet in height. It is compact with dense foliage and bears well in hot, dry areas. 'Fairchild' is similar but will tolerate milder climates. Mango fruit is widely used, eaten raw or cooked and made into preserves and chutneys. Many parts of the tree are used medicinally. Its wood is used to make furniture, bowls, and plywood. It figures heavily in Hindu religious ceremonies and texts. The mango flower is invoked as one of the five arrows of Kama Deva, the Hindu Dan Cupid. The Buddha was given a mango grove to provide shaded repose. In addition to its fruit, the mango tree is an excellent screening tree or may be used as a windbreak.

Mango is a moderate to rapid grower, depending on variety. All are worthwhile for the edible xeriscape. (height 120 feet/canopy 100 feet)

Phytolacca dioica

OMBU TREE (P)

ZONE 4 H(HPWRA)

Phytolaccaceaae (Pokeweed Family)

Native to dry grasslands of Argentina, Brazil, and Uruguay, this tree will reach 60 feet in height with an equal spread. It develops a water-storing, broad base from which several trunks may grow. It is heat, drought, and wind tolerant and fire resistant. Note that all parts of the plant are poisonous. Plant it in full sun in a well-drained soil. Use it as an attention-getting or conversation piece, an accent tree, windbreak, or screen. Clearly not a tree for the small residence, this tough tree would be at home in the rough part of a golf course in a hot, dry area. Plant it where contact with its sap is unlikely. It needs full sun and a well-drained soil. It is a slow grower. (height 60 feet/canopy 60 feet)

Platymiscium pinnatum

ROBLE, QUIRA

ZONE 3

Fabaceae (Bean Family)

A rapid grower, this tree from Central America and northern South America will reach 65 feet in height. It is deciduous in spring and displays masses of fragrant golden yellow flowers during that period. It has a low tolerance to salt but a moderate to high drought and wind tolerance. It thrives in good, well-drained soils in full sun. Use it in park and resort plantings and school landscapes where its bright spring color will make a strong accent. Its fine-grained, durable, reddish heartwood is prized for use in cabinetry and furniture making. (height 65 feet/canopy 65 feet)

Polyalthia longifolia var. *pendula*

WEEPING DEBDAR

ZONE 3

Annonaceae (Custard Apple Family)

Southern India and Sri Lanka are the original homes of this slow-growing evergreen tree that will reach 60 feet in height. It produces a narrow crown of slender, dark green leaves with wavy margins. It is a fine ornamental tree, providing a strong vertical accent in the garden. Flowers are inconspicuous. It is heat and wind tolerant and moderately tolerant of drought. Plant it in full sun in a well-drained soil. (height 60 feet/canopy 10 feet)

Pterocarpus indicus

NARRA, BURMESE ROSEWOOD, BLOODWOOD N

ZONE 2 L(HAWAI'I)

Fabaceae (Bean Family)

P. indicus forma *echinatus*

Showing moderate drought, heat, and wind tolerance but low tolerance to salt, this useful tree from Southeast Asia and the Philippines grows rapidly to a height of 80 feet. It has a weeping habit, forming a stately silhouette. Its flowers are ornamental, yellow, and produced in spring. They are followed by papery seed pods that do not present a maintenance problem and are used in dry arrangements and wreaths. A nitrogen fixer, it will thrive in almost any well-drained soil. Narra is an excellent choice for parks, campuses, and other large properties. A similar *Pterocarpus, P. indicus* forma *echinatus,* differs in having somewhat spiny, papery seed pods. (height 80 feet/canopy 80 feet)

Pterocarpus marsupium

EAST INDIAN KINO N

ZONE 3

Fabaceae (Bean Family)

Native to seasonally dry areas of India and Sri Lanka, this species grows at a rather slow rate to 60 feet or more in height, forming a rather columnar shape. Conspicuous flowers, profusely borne, are yellow and followed by papery seed pods that may be used in dry arrangements and wreaths. The name "kino" comes from the bloodred sap (a gum resin) when the tree is cut. In addition to its landscape usefulness, it provides a fine wood for furniture making and carving. It shows moderate drought tolerance and good tolerance of heat and wind. It may be used in groves and as a screening tree, as well as for its shade. It is briefly deciduous. (height 60 feet/canopy 30 feet)

Reynoldsia sandwicensis

'OHE, 'OHE MAKAI

ZONE 4

Aralicaeae (Genseng Family)

Found in dry forests on all the main islands except Kaua'i and Kaho'olawe, this endemic Hawaiian species will grow to 60 feet in height. In the wild it is deciduous during the dry season, but with a little watering it will retain its ornamental foliage all year and develop a fairly straight trunk. It displays a beautiful, smooth white bark. Its canopy is spreading, providing good shade. Small greenish yellow flowers are followed by attractive dark purple fruit. Plant it in full sun in any well-drained soil. Do not plant it where its purple fruits will stain pavement. This is true of other fruit trees. Use it as a featured tree or an accent, or for shade. It is beautiful when uplighted at night. It is moderately drought, heat, and wind tolerant. It is a moderate grower. (height 60 feet/canopy 60 feet)

Samanea saman

MONKEYPOD N

ZONE 3

Fabaceae (Bean Family)

Native to the extensive area from Mexico south to Peru and Brazil, this useful tree will reach 75 feet in height. It is a rapid grower in most well-drained soils. It is wind, heat, and moderately drought tolerant and is valued for planting in parks, golf courses, and other large spaces. It provides excellent shade. Its pink flowers in spring are highly attractive. Monkeypod is used by the wood carver to produce bowls and trays. (height 75 feet/canopy 150 feet)

Sterculia foetida

JAVA OLIVE

ZONE 3

Sterculaceae (Cacao Family)

A moderate grower reaching as much as 115 feet in height, this unusual deciduous tree from seasonally dry areas of Old World tropics bears handsome foliage and highly decorative, large, red seed pods containing large blue black seeds. These are collected by the flower arranger. Unhappily, the showy red flowers, which appear in the spring, carry an unpleasant odor that is detectable from quite a distance. Happily, however, the flowering period is brief. It is sometimes called the skunk tree for the above reason. The tree should be planted downwind in full sun in a well-drained soil. It is heat, wind, and moderately drought tolerant but shows a low tolerance to salt. Use it as an accent in a large property, being careful to plant it where neighbors will not be offended. (height 115 feet/canopy 100 feet)

Swietenia mahagoni

CUBAN MAHOGANY

ZONE 3

Meliaceae (Mahogany Family)

Growing rather rapidly to 75 feet in height, this famous tree is prized for the excellent quality of its wood, which was used extensively by fine English cabinetmakers. Native to the extreme southern part of Florida, the Bahamas and into Cuba, Jamaica, and Hispaniola, it is wind and heat tolerant and moderately drought tolerant. It is a fine shade tree for large spaces, parks, schools, and golf courses. Honolulu boasts an attractive avenue of Cuban Mahogany trees planted in the early 1900s. (height 75 feet/canopy 50 feet)

Tabebuia chrysea

GOLDEN TRUMPET TREE

ZONE 3

Bignoniaceae (Catalpa Family)

Native to dry parts of northern Venezuela and Columbia, this species grows to 60 feet in height, bearing masses of golden yellow flowers. It is a moderate grower and requires sun and a well-drained soil. Use it in the xeriscape as a spectacular color accent, screen, or windbreak and for its shade. It is evergreen and shows tolerance of heat and wind and moderate tolerance of drought. (height 60 feet/canopy 45 feet)

Tabebuia donnell-smithii

GOLD TREE

ZONE 3

Bignoniaceae (Catalpa Family)

From Mexico and Guatemala, this species forms a tall, somewhat columnar tree reaching 100 feet in height. Spectacular masses of brilliant yellow flowers are borne erratically in winter or spring, and some specimens will flower a second time in fall. Sometimes only half of the tree will flower, while the balance is bare. It is deciduous and a moderate grower, thriving in almost any well-drained soil. It shows moderate drought and wind tolerance but low salt tolerance. It is an almost unmatched color accent. Mature flowering specimens may be seen from a great distance. (height 100 feet/canopy 50 feet)

Tabebuia heterophylla

PINK TECOMA

ZONE 3

Bignoniaceae (Catalpa Family)

Native to the West Indies, this semi-evergreen species will reach 60 feet in height, bearing masses of flowers in the spring, although flowering may occur occasionally during the year. It produces a columnar silhouette and is valuable in the xeriscape as a showy vertical accent, windbreak, or screen. Flower color may vary from light to dark pink to white, while variations in foliage color range from light to dark green. Even branching habit may vary. To assure planting the qualities the gardener prefers, it must be propagated by cuttings or air layers. It displays moderate heat, wind, drought, and salt tolerance. (height 60 feet/canopy 40 feet)

Tabebuia impetiginosa

AMAPA

ZONE 3

Bignoniaceae (Catalpa Family)

A large trumpet tree from the area ranging from Mexico south to Argentina, this species will rather slowly reach 90 feet in height. When in flower, usually in winter when the tree is partially deciduous, it presents a spectacular sight. Flowers, produced in dense spherical clusters, are highly variable as to color and range from pale pink to a strong reddish pink. Its hard wood is tough and durable and is of commercial value. It shows good tolerance of heat and wind and is moderately drought tolerant. It performs best in a well-drained soil. Use it as a great color accent or as a tall screen or hedge. (height 90 feet/canopy 45 feet)

Tabebuia ochracea subsp. *neochrysantha*

YELLOW TRUMPET TREE

ZONE 2

Bignonicaeae (Catalpa Family)

Frequently flowering several times during the spring and summer months, this is a spectacular flowering accent, growing to as much as 60 feet in height. It develops a rounded canopy and is deciduous for a short period. It is native to Central and South America, and although preferring a moist soil, it will tolerate limited drought. In the wild it is listed as occurring in both arid and high rainfall environments. Experiment. It is a moderately fast grower. This is an excellent color accent and makes a dramatic avenue or mass planting. Roots are not aggressive. It provides a bright color accent. (height 60 feet/canopy 45 feet)

Tabebuia rosea

ROSY TECOMA TREE

ZONE 3 EVALUATE

Bignoniaceae (Catalpa Family)

A showy deciduous tree native to Mexico and south to Venezuela and Ecuador, this species will attain 75 feet in height, displaying clusters of flowers varying in color from dark pink to almost white. It is an important timber tree in Central America and is the national tree of El Salvador. Use it as an important color accent and for its shade. It does best in an open, well-drained soil. It has fair tolerance to heat and wind and moderate tolerance to drought. Use it as a striking color accent, for an avenue planting or as a tall colorful screen. It is a moderate grower. (height 75 feet/canopy 50 feet)

Tamarindus indica

TAMARIND NP

ZONE 4

Fabaceae (Bean Family)

Native to dry savannas of tropical Africa, the tamarind was introduced into India in ancient times by Arab traders. It became so widespread there that both its scientific and common names are associated with India. Marco Polo noted tamarind in AD 1298, and it became popular in Medieval Europe. Prized for the pleasant, sweet-sour pulp in its fruit, it has long been used to concoct beverages, chutney, and curries and in ice cream. It is high in vitamins and minerals and has medicinal applications. The dark, purplish brown heartwood is used in furniture making, paneling, and toy making. It is termite resistant. Attractive flowers are small, yellow with pink veins, and are borne abundantly. Young leaves, flowers, and pods are used to flavor rice, meat, and fish dishes. Foliage is fine, almost lacy. Tamarind is drought and wind tolerant and moderately salt tolerant. It is not particular as to soil as long as it is well drained, and it will reach 80 feet in height. Use it in large xeriscapes as a focal point, a tall screen or hedge, or simply as a beautiful shade tree. It is a rather slow grower. (height 80 feet/canopy 60 feet)

Tipuana tipu

ROSEWOOD, TIPU, PRIDE OF BOLIVIA N

ZONE 4

Fabaceae (Bean Family)

Producing masses of yellow flowers with a reddish throat in spring and summer, this tree will reach 60 feet in height, developing a broad canopy and light shade. It is heat, wind, and drought tolerant and can be used as a color accent and shade tree in the large xeriscape. It is native to the broad area from southern Brazil and northern Argentina to Bolivia. It thrives in almost any soil as long as it is well drained and must have full sun. It is a moderate grower. (height 60 feet/canopy 60 feet)

Vitex doniana

BLACK PLUM, MERU OAK N

ZONE 4

Verbenaceae (Verbena Family)

Found throughout both moist and dry parts of tropical Africa, this tree will reach 75 feet in height. One reference states 95 feet in height. It is a moderate grower. It will thrive in any open, well-drained soil. It is heat, wind, and drought tolerant. Its small flowers, rather inconspicuous, are white with one violet petal. It bears edible, sweet fruit. Its wood is used as a base for fine plywood. Parts of the tree have medicinal applications. It develops a light canopy, permitting the growth of other xerophytes below. Use it for its shade or as a screen. It is evergreen. (height 75 feet/canopy 50 feet)

VINES

Allamanda blanchetii

PURPLE ALLAMANDA **P**

ZONE 2 EVALUATE

Apocynaceae (Dogbane Family)

Native to Brazil, this is a rather sprawling, vinelike plant reaching 6 feet in height. It produces its attractive flowers in spring and summer. It has moderate drought and wind tolerance but is not tolerant of salts. Use it as an evergreen ground or deep bank cover or as a colorful, informal hedge. It is a moderate grower.

Allamanda cathartica

COMMON ALLAMANDA **P**

ZONE 2 EVALUATE

Apocynaceae (Dogbane Family)

'Hendersonii'

'Stansell's Double'

A tropical American vine climbing to 50 feet and bearing fragrant yellow flowers all year, this species likes a loamy soil with good drainage. Usually seen in well-watered places, it is, however, moderately tolerant of drought and wind but is not tolerant of salt. A common cultivar is Henderson's Allamanda (*A. cathartica* 'Hendersonii'). It will grow on a fence or pergola and can be used to cover a bank. Light pruning will make it bushier. Another cultivar, 'Stansell's Double', produces double flowers. They are fairly rapid growers and may become invasive.

Aloe ciliaris

CLIMBING ALOE

ZONE 4

Aloeaceae (Aloe Family)

Native to South Africa, this species forms sprawling vinelike stems to 10 feet or more in length that form a low, wide-spreading ground or bank cover. It may also be seen as a climber on large shrubs and tree trunks. Its scarlet red flowers are tipped with yellowish green, providing a colorful accent in the water-conserving garden. It is a good erosion control plant for a slope or is valued in the rockery. Full sun and well-drained soil produce the best plants. It is a slow grower.

Argyreia nervosa

WOOLY MORNING GLORY,
BABY WOOD ROSE **P**

ZONE 3

Convolulaceae
(Morning Glory Family)

Performing best in a rich, well-drained soil in full sun, this Indian vine will rapidly climb to 40 feet. It shows moderate tolerance to wind and drought but not to salt. Use it to cover a pergola or high fence, or it can make a good, steep bank cover. Its dried fruit clusters are used in arrangements or in a corsage.

Bauhinia galpinii

RED BAUHINIA,
NASTURTIUM BAUHINIA **N**

ZONE 3

Fabaceae (Bean Family)

An evergreen vine from tropical Africa climbing to 40 feet, this species is usually seen as a heavily pruned shrub at 15 feet in height. Its showy flowers are produced from spring into fall and are variously colored brick red to a strong yellow red. It grows best in a good, well-drained soil in full sun. It is an excellent pergola or strong trellis cover and can be used to make a tall bank cover or, if pruned, a large, colorful accent shrub. It shows moderate drought and wind tolerance but is not tolerant of salts. It is a fairly rapid grower.

Beaumontia multiflora

EASTER LILY VINE

ZONE 2

Apocynaceae (Dogbane Family)

Growing best in full sun in a good, well-drained soil, this is a slow-growing woody vine from India climbing 25 feet. Its large white flowers have a spicy fragrance. Plant it on a pergola or trellis where the fragrance of its flowers may be enjoyed. It has moderate drought tolerance. It performs best given full sun and an open, well-drained soil.

Bonamia menziesii

BONAMIA VINE

ZONE 4

Convolvulaceae
(Morning Glory Family)

An endemic and endangered vine, this attractive species is useful in the xeriscape. It climbs to 30 feet, bearing flowers of a greenish white color. It is an excellent pergola or trellis cover or can be used to grow on a fence. Full sun and excellent drainage are required. It is wind, heat, and drought tolerant. It is a moderate grower.

Bougainvillea 'California Gold'

Bougainvillea spectabilis

BOUGAINVILLEA **T +**

ZONE 3

Nyctaginaceae (Four O'Clock Family)

'California Gold'

'Carmencita'

'Mary Palmer'

'Miss Manila'

'Raspberry Ice'

One of nature's most popular productions, bougainvillea is planted worldwide in the tropics and subtropics for the spectacular color of its hybrids. The species, a woody Brazilian native growing to 50 feet, is purple, while hybrids vary in color from white through pink, orange, gold, purple, lavender, and bicolors. There are double-flowered hybrids as well as those with variegated foliage. Pruning and regular feeding enhance flowering, which occurs on new growth. They are moderately drought, salt, and wind tolerant and perform well in a variety of soils with good drainage. Bougainvillea is used as a large pergola cover, deep bank cover, or as a protective, high hedge. It can be espaliered on a high wall or trained along eaves. It forms an effective barrier hedge. It is also a popular potted or tubbed specimen, in which case almost constant pruning and feeding are required. If not sufficiently watered, all foliage drops. Increase applied water. Bougainvilleas are at the top of the list for long-term flower color. They are rapid growers.

Canavalia cathartica

MAUNALOA

ZONE 4

Fabaceae (Bean Family)

A rapid-growing tropical vine, Maunaloa is native to the area ranging from East Africa to India and the Ryukyu Islands and through Malesia to Polynesia. Naturalized in Hawai'i, it is found in dry areas and has become a favored flower for creating the prized maunaloa lei. Flower color may vary from pale pink to rose. It is drought tolerant. Use it to cover a fence, pergola, or trellis where its flowers may be harvested. Full sun and a well-drained soil are required.

Canavalia galeata

ʻĀWIKIWIKI

ZONE 4

Fabaceae (Bean Family)

An endemic vine, this native species is found only on Oʻahu at low, fairly dry elevations, bearing dark purple, pea-shaped flowers heightened by a white spot. It is favored by the lei maker. It is a rapid grower and will quickly cover a fence or trellis. Full sun and a well-drained soil will produce best results. It is moderately to possibly fully drought tolerant. Other moderately drought tolerant endemic species (not figured) are *C. hawaiiensis,* with dark carmine flowers that have a white dot at the base from dry areas on the Big Island; and *C. pubescens,* found in dry areas on Niʻihau, Kauaʻi, Lanaʻi, and East Maui. It has dark red to pink flowers.

Cissus rhombifolia

GRAPE IVY

ZONE 4

Vitaceae (Grape Family)

Native to the Caribbean and south to Brazil, this somewhat succulent, vining grape ivy will reach 10 feet in height and may be used as a fence cover or as a ground cover mounding to a foot or so in depth. It is also useful as an erosion control species. It is heat, wind, and drought tolerant and grows best in a well-drained soil in full sun. It is a rather rapid grower.

Cissus rotundifolia

ARABIAN WAX CISSUS

ZONE 4 H(HPWRA)

Vitaceae (Grape Family)

Native to East Africa and southern Arabia, this vine bears succulent foliage and is drought, heat, and wind tolerant and shows moderate tolerance of salt. Use it to cover a chain-link fence, trellis, or arbor or as an espalier on a wall. Like most grapes, it has tendrils to hold it in place. It will reach 30 feet in length. Plant it in full sun in a well-drained soil. It is a moderately rapid grower.

Congea griffithiana

SHOWER OF ORCHIDS VINE,
PINK SANDPAPER VINE

ZONE 2

Verbenaceae (Verbena Family)

Growing best in full sun in a good soil with drainage, this sprawling Malaysian vine will reach 50 feet in height. It produces masses of pink flowers in winter or late spring and is used as a massed planting for large spaces or for covering a bank or trained over a large pergola or trellis. It has moderate drought and wind tolerance but no salt tolerance. It is a fairly rapid grower.

Congea tomentosa

CONGEA

ZONE 2

Verbenaceae (Verbena Family)

Possibly of Malaysian origin, this is very similar to the shower of orchids vine but differs visually in displaying branches, inflorescences, and the underside of leaves covered with white woolly hairs. Lilac-colored bracts of the inflorescence provide interesting color. With lighting it becomes a bright focal point in the nightscape. It is a moderate grower.

Ficus pumila

CREEPING FIG, CLIMBING FIG

ZONE 3

Moraceae (Mulberry Family)

'Variegata'

Capable of climbing to 100 feet, this is a tough, vigorous vine from Southeast Asia that attaches itself to almost any surface with special roots. Do not plant it where it can climb a wooden structure, as the roots penetrate the wood and eventually ruin it. It is used to cover hollow tile or rock walls and concrete surfaces. Several cultivars are available with dwarf or variegated foliage. It grows in any well-drained soil and shows moderate tolerance to drought. It prefers full sun but will grow satisfactorily in light shade. It is a fairly rapid grower.

Ipomoea cairica: Koali (see "Ground Covers")

Ipomoea pes-caprae subsp. *brasiliensis:* Pohuehue (see "Ground Covers")

Jasminum laurifolium forma *nitidum*

CONFEDERATE JASMINE, ANGEL WING JASMINE

ZONE 2

Oleaceae (Olive Family)

A moderately fast growing, climbing Indian species, this vine will reach 20 feet in height. Its fragrant flowers may be seen most of the year. It is a good choice for a small trellis or pergola or may be trained along the top of a fence to increase screening. It needs full sun and a well-drained soil. It has moderate drought and salt tolerance.

Jasminum multiflorum

STAR JASMINE

ZONE 2

Oleaceae (Olive Family)

An evergreen twining vine from India growing to 20 or 30 feet in height, this species produces flowers much of the year. Performing best in full sun in a well-drained soil, star jasmine displays moderate tolerance of drought and salt. Use it as a fence cover, on a pergola or trellis, or trained along the eaves of a house. It is a rapid grower.

Lonicera x *heckrottii*

GOLDEN FLAME HONEYSUCKLE, PINK HONEYSUCKLE

ZONE 2

Caprifoliaceae (Honeysuckle Family)

Thriving in most well-drained soils and planted in full sun or light shade, this evergreen hybrid vine will reach 15 feet in height. It shows moderate tolerance to drought and salt and excellent wind tolerance. It can be used as a sprawling ground or bank cover or on a small trellis or pergola. It is a rapid grower.

Marsdenia floribunda

STEPHANOTIS, MADAGASCAR JASMINE

ZONE 3

Asclepiadaceae (Milk Weed Family)

Prized for its many clusters of highly fragrant, white flowers produced in late winter, spring, and into summer, stephanotis is a moderately fast growing, evergreen twining vine reaching 30 feet in length. It is native to Madagascar. Plant this vine in full sun in a well-drained soil and provide the means for its growth: a trellis, chain-link fence, or a band of mesh strung along an eave of the house. It is heat and wind tolerant and will tolerate moderate drought.

Passiflora vitifolia

RED PASSION FLOWER

ZONE 2 H(HPWRA)

Passifloraceae (Passion Flower Family)

Bearing its bright red flowers much of the year, this fast-growing evergreen vine from Venezuela to Bolivia is a good choice for covering a pergola, arbor, or trellis. Plant it in full sun in a well-drained soil. It has low salt tolerance but moderate tolerance to drought and wind.

Petrea volubilis

SANDPAPER VINE, PURPLE WREATH

ZONE 3

Verbenaceae (Verbena Family)

'Albiflora'

Climbing to 35 feet or permitted to form a large mound, this tough vine from the West Indies, Central America, and Mexico shows moderate drought and salt tolerance. Its flowers are borne on branch tips during late winter and spring and may be used by the lei maker. There is a white-flowered cultivar, *P. volubilis* 'Albiflora'. The name "sandpaper vine" is derived from the rough, sandpapery texture of its leaves. Use it as a tall, mass bank cover or trained along roof eaves, on a pergola, or arbor. It is a moderate grower.

Podranea ricasoliana

PINK TRUMPET VINE

ZONE 2 EVALUATE

Bignoniaceae (Catalpa Family)

Planted in a good, well-drained soil in full sun, this South African vine will climb 20 feet on a pergola, trellis, or arbor. Its flower clusters, varying from pure white to deep pink, appear much of the year. It shows moderate drought, wind, and salt tolerance. It is a moderate grower.

Pseudogynoxys chenopodioides

MEXICAN FLAME VINE,
ORANGE-FLOWERED SENECIO **P**

ZONE 3 H(HPWRA)

Asteraceae (Sunflower Family)

Reaching 40 feet in height, this Mexican evergreen vine bears masses of brilliant orange flowers from spring to fall. It grows readily in full sun or light shade in most soils, providing a bright color accent as a fence cover, on an arbor or pergola, or on a trellis; unsupported, it will make a good bank cover. It has good drought and wind tolerance and moderate salt tolerance. It is a rapid grower.

Quisqualis indica

RANGOON CREEPER

ZONE 2

Combretceae (Combretum Family)

This woody evergreen vine from Myanmar to New Guinea will climb to 30 feet, showing its clusters of fragrant flowers most of the year. It tolerates a wide range of soils and is moderately drought and wind tolerant. It does not tolerate salt. Plant the Rangoon creeper in a contained area—for example, in a planting space in a paved area or between a paved walkway and a building—as it tends to produce suckers and can become invasive. It makes a good cover for an arbor or large pergola or on a vertical trellis against a tall wall. It is a fairly rapid grower.

Bougainvillea spectabilis

Saritaea magnifica

PURPLE BIGNONIA

ZONE 2 EVALUATE

Bignoniaceae (Catalpa Family)

Climbing rapidly to 60 feet, this woody Colombian species bears its flowers in the fall. It is an excellent arbor or vertical trellis vine. Preferring full sun in a well-drained soil, purple bignonia displays moderate drought and wind tolerance but no salt tolerance.

Stigmaphyllon ciliatum

BRAZILIAN GOLD VINE

ZONE 2

Malpighiaceae (Malpighia Family)

In addition to its showy flowers, this vine is popular for its highly ornamental fringed leaves. It is native to parts of eastern Central and South America and south to Uruguay. Use it to cover a small arbor or doorway trellis and as an interesting accent. It has only moderate tolerance of heat, wind, and drought and is a moderate grower.

Stigmaphyllon floribundum

ORCHID VINE

ZONE 2

Malpighiaceae (Malpighia Family)

Native to the Caribbean and Atlantic coasts of Central and South America to Uruguay, this slender vine may reach 20 feet in length. It is used to cover fences, over a small arbor, or even along roof eaves. It shows moderate wind, heat, and drought tolerance. Plant it in full sun in a well-drained soil. It is a moderate grower.

Trachelospermum jasminoides

CONFEDERATE JASMINE, STAR JASMINE, MAILE HAOLE

ZONE 3

Apocynaceae (Dogbane Family)

'Variegatum'

A Chinese evergreen vine climbing to 30 feet, this species bears masses of white, extremely fragrant flowers from spring to summer. It prefers full sun and a loamy soil with good drainage. It performs best at cooler elevations but will flower at sea level. It is frequently used as a tumbling ground or bank cover but is excellent for covering a trellis or archway, or trained along roof eaves. Foliage may be used by the lei maker as a maile substitute. It has moderate drought and salt tolerance. *T. jasminoiodes* 'Variegatum' produces small green and white variegated foliage tinged with red. It is a moderate grower.

Tristellateia australasiae

BAGNIT VINE, SPRAY OF GOLD

ZONE 3

Malphigiaceae (Malphigia Family)

Native to the area from Southeast Asia to New Caledonia, this vine will grow to 20 feet, producing its clusters of bright yellow flowers much of the year. It is a good vine for roof eaves, for covering a trellis or arbor, or as an espalier given a vertical construction to support it. It displays moderate wind, salt, and drought tolerance and is a moderate grower.

Vitis x *prolifera* 'Isabella'

ISABELLA GRAPE

ZONE 3

Vitaceae (Grape Family)

Possibly native to hot, dry areas of the eastern Mediterranean, the Old World grape has been hybridized with the American grape to produce a variety of types used to produce wine, raisins, currants, jelly and jam, and for table grapes. In Hawai'i the vine is almost evergreen, producing grapes most of the year, with the high point of production during the warm months. It makes an ornamental fence or eave cover but is best grown over an arbor, where its dappled shade may be enjoyed on a hot day and its pendant fruit clusters can be viewed and picked from below. Heavy pruning during the winter months will keep the vine in control. It is rapid growing and must have full sun and a well-drained soil. It is possible that the Isabella grape was introduced to Hawai'i by Don Marin. It is known that as early as 1820, Marin was producing wine and brandy at his vineyard on the *makai* side of what is now known as Vineyard Boulevard, across from Foster Botanical Garden. The grape is of great cultural importance. Known in classical times as "the vine," it figures in myths and legends: the grape is the symbol of Bacchus, the Greek God of Conviviality. The vine is also the symbol of the Jewish nation and the emblem of Christianity, Jesus Christ being referred to as "The Vine." It is heat and wind tolerant and moderately drought tolerant.

BROMELIADS

BROMELIAD NOTES

Listed are but a few of the many bromeliads currently available to the gardener. For the xeriscape, we list only major types that we feel make a significant landscape impact. All are in the Bromeliaceae, the pineapple family, and all are native to the American tropics. They are slow growers. There are several good bromeliad publications listed in the suggested readings.

Aechmea blanchetiana

ORANGE BLANCHETIANA, LEMON BLANCHETIANA T

ZONE 3

Bromeliaceae (Pineapple Family)

Growing to 3 feet in height, both the orange- and lemon-leaved forms of this bromel provide major color accents in the garden. Full sun intensifies the color. It requires a very porous soil and displays good heat and salt wind tolerance and moderate drought tolerance.

Alcantarea imperialis

IMPERIAL BROMELIAD

ZONE 3

Bromeiaceae (Pineapple Family)

A large Brazilian species forming a rosette of leaves up to 5 feet in length or 10 feet across, this bromel is one of the finest color accents available to the xeriscaper. Plant it in full sun for developing deep color. It will grow both terrestrially (in a well-drained medium) or epiphytically. It tolerates heat, drought, wind, and moderate exposure to salt air.

Ananas comosus var. *variegatus*

VARIEGATED PINEAPPLE T

ZONE 3

Bromeliaceae (Pineapple Family)

Smaller than commercial pineapple, this is a color form showing leaves longitudinally striped with whitish cream and green. When grown in full sun, the entire plant including its small fruit turns pinkish. In the landscape, use it as a strong color accent among smaller, less colorful plants. It shows good heat and wind tolerance and moderate tolerance of drought and salt.

Alcantarea imperialis

Tillandsia rothii

ROTH'S TILLANDSIA

ZONE 3

Bromeliaceae (Pineapple Family)

Moderately drought tolerant, this slow-growing bromel from the west coast of Mexico produces a spectacular, long-lasting inflorescence rising well above the plant. It consists of many pale green flowers emerging from a mass of cherry, green, and yellow bracts. The plant forms a colorful rosette of foliage 18 inches across. Do not permit water to stand in leaf axils. Use it in the xeriscape as a color accent. Plant it where its intricate color may be appreciated. It is epiphytic.

Tillandsia tectorum

ROOF TILLANDSIA

ZONE 4

Bromeliaceae (Pineapple Family)

A highly variable species from Ecuador and Peru, the roof tillandsia may produce a rosette of slender, silvery white leaves up to 10 inches across capped by a long-lasting inflorescence showing strong pink, purple and aster violet parts. Place it in full sun. It is a slow grower. In the wild it is found in colonies on rock outcroppings and cliff faces. In the dry garden, use it in similar situations as a bright foliar and flower accent. It may also be used as an epiphyte but placed where it receives full light.

Tillandsia xerographica

DESERT TILLANDSIA

ZONE 4

Bromeliaceae (Pineapple Family)

A popular species from very dry parts of southern Mexico to Guatemala and El Salvador, this tillandsia forms a rosette of foliage 10 inches across and as high. Its highly colorful inflorescence, carried well above the silvery gray foliage, displays salmon to clear pink, yellow, and mauve parts. The flowering period may last for several months. It is a slow grower requiring full sun and perfect drainage. It is heat, wind, and drought tolerant. Use this spectacular plant in the rockery, especially against dark lava or red cinder, as an accent planted close at hand to permit enjoyment of its flowers.

CYCADS

CYCAD NOTES

Cycads are primitive members of the plant kingdom and were widespread during the Mesozoic era. There are three plant families composing the cycads. These are Cycadaceae, Stangeriaceae, and Zamiaceae. Some are from moist environments while others, of interest to the xeriscaper, are native to dry areas. We list only a few below but list several good publications in the suggested readings.

Cycas circinalis

ZONE 2

SAGO PALM, QUEEN SAGO P T

Cycadaceae (Cycad Family)

Native to Asian tropics, this cycad grows slowly to 25 feet in height, producing many offsets that may develop into secondary trunks. Most gardeners remove excess offsets to retain a pleasing form. It shows moderate tolerance to drought, wind, and salt. Its leaves, up to 8 feet in length, are used in large arrangements. Use it as a shade tree. Fortunately it is available in large sizes that transplant easily. Planted as a small specimen, space must be given for the development of its broad canopy, in which situation it makes a dark green background for other plants.

Cycas revoluta

SAGO PALM, JAPANESE SAGO PALM, KING SAGO P T

ZONE 2

Cycadaceae (Cycad Family)

Native to southern Japan, this is a slow-growing cycad reaching 10 feet in height producing many offshoots that may develop into secondary trunks. Most gardeners remove excess offsets to retain a pleasing form. Its long, glossy, dark green leaves, up to 6 feet in length, are popular for making arrangements. It grows easily in any well-drained soil in both sun and light shade. Use it in the xeriscape as an accent among light green or gray green plants or as a background plant. It shows moderate tolerance to drought and good tolerance of heat, wind, and salt. Large specimens are available and transplant easily.

Encephalartos eugene-maraisii

MARAIS CYCAD

ZONE 4

Zamiaceae (Coontie Family)

This cycad from the Northern Province of South Africa is noted for the silvery gray color of its foliage, which may grow to 4.5 feet in length. A slow grower, it may eventually reach 6 feet in height, producing a few suckers from its base. Use this rather spectacular plant as a bright accent among dark-leaved plants or among dark rocks. It fares best in full sun in a very well drained soil. It is heat, wind, and drought tolerant. Use it in the nightscape.

Encephalartos horridus

EASTERN CAPE BLUE CYCAD S

ZONE 3

Zamiaceae (Coontie Family)

Named for its easily avoided, silver blue, prickly leaves, this endangered South African cycad will reach 3 feet in height with an equal spread. It is a slow grower requiring a well-drained soil and full sun. It is heat, wind, and moderately drought tolerant. It provides a striking color accent when planted among dark rocks or surrounded by dark-foliaged plants and is brilliant when uplighted in the night garden.

Encephalartos lehmannii

KAROO CYCAD **S**

ZONE 3

Zamiaceae (Coontie Family)

With its leaf color enhanced by full sun, this cycad presents a striking silvery blue accent for the xeriscape. It requires good drainage and is heat, salt, and wind tolerant and moderately tolerant of drought. A slow grower, its trunk may reach 6 feet in height. It forms a clump. It is native to semiarid regions of South Africa. Plant it where its sharp-pointed leaves do not touch people. It is beautiful in the night garden with uplighting.

Encephalartos trispinosus

THREE SPINE ENCEPHALARTOS **S**

ZONE 3

Zamiaceae (Coontie Family)

This colorful cycad is native to dry forests and scrublands of South Africa. Its bright blue green foliage provides a strong accent in the xeriscape, especially in the night garden with uplighting. It will grow slowly to 3 feet in height, forming a small clump. It is heat, wind, and drought tolerant and possibly withstands salt wind. In addition to its accenting quality, it also makes a fine bank cover or in a rockery. Plant it where its sharp spines do not touch people.

Macrozamia macdonnellii

MACDONNELL RANGE CYCAD

ZONE 3

Zamiaceae (Coontie Family)

Native to semiarid areas of central Australia, this cycad slowly develops a trunk reaching 10 feet in height. Its long, young leaves are powdery blue in color, maturing to a blue green. It must have full sun, excellent drainage, and lots of air movement. It is heat, wind, and moderately drought tolerant. Use this as a major accent both for its form as well as its foliage color. Uplighting in the night garden produces a dramatic effect.

Macrozamia moorei

MOORE CYCAD

ZONE 3

Zamiaceae (Coontie Family)

Another Australian cycad, this is a large species with a canopy or leaf spread of 16 feet or more. It must have space. It is considered a rapid grower, and handsome specimens may be realized with a trunk 10 feet in height within several years. In the wild, old specimens have been seen up to 30 feet in height. One author considers this species "the most beautiful and majestic of all the species in the genus." It is heat and wind tolerant and moderately drought tolerant. Leaflets are tipped with a sharp spine and should be planted where unwary people contact is unlikely. Use it as a colorful accent as a background for shrubby plantings with dark green foliage. A stunning accent in the xeriscape with nightlighting.

Zamia furfuracea

CARDBOARD PALM **P**

ZONE 4

Zamiaceae (Coontie Family)

Native to coastal sand dunes and offshore islands in Vera Cruz on Mexico's southeastern Gulf Coast, this popular cycad develops into a large clump up to 8 feet in height and 16 feet across. It is a very tough species, enduring wind, heat, drought, salt air, and neglect. In the wild it may be found growing just above the high tide line. Growing best in full sun, it will also grow in light shade and in almost any well-drained soil as well as sandy soils. Use it as a large bank cover, an accent, or as a tall background or screen.

Zamia integrifolia

FLORIDA ARROWROOT, INDIAN BREADROOT, COONTIE

ZONE 2

Zamiaceae (Coontie Family)

Extensive harvesting of this species in the nineteenth century was done for processing into Florida arrowroot, used in the making of bread and crackers. This severely reduced wild populations. Its popularity for landscaping, however, has resulted in heavy propagation, and plantings are now commonly seen even in public spaces. Its medium to dark green, glossy leaves vary from 18 to 36 inches in length, usually forming, in time, a multiheaded clump of that height and twice that in spread. Its habitat includes dry oak or pinelands, commonly in sand. On the Gulf Coast, it is sometimes associated with shell deposits left by early Indians, for whom it was a food staple. It is native to extreme southeastern Georgia and throughout Florida, including the Florida Keys. One reference indicates that its wider habitat includes the Bahamas, central Cuba, and the Dominican Republic. Plant this species in filtered light in a sandy or other well-drained soil. It is heat, wind, and moderately drought and salt air tolerant. It may be used as a low border, ground cover, or bank cover.

APPENDIX A

PLANTS FOR HEDGES, SCREENS, AND WINDBREAKS

ZONE 2

Acalypha wilkesiana
Breynia disticha 'Roseopicta'
Hibiscus rosa-sinensis
H. schizopetalus
Polyscias guilfoylei
Schefflera arboricola
Tabernaemontana divaricata
Tamarix chinensis
Thunbergia erecta

ZONE 3

Brunfelsia americana
Carissa macrocarpa
Codiaeum variegatum
Duranta erecta
Ficus microcapa var. *crassifolia*
Gardenia volkensii
Holmskioldia sanguinea
Lysiloma thornberi
Murraya paniculata
Pittosporum tobira
Senna polyphylla

ZONE 4

Aloe africana
A. castanea
A. excelsa
A. kedongensis
A. marlothii
A. suzannae
A. volkensii
Bontia daphnoides
Calotropis gigantea
C. procera
Dodonaea viscosa (H)
Euphorbia antiquorum
E. lactea
E. neriifolia
E. royleana
Kalanchoe hillebrandtii
Nerium oleander
Plumeria stenopetala
P. stenophylla
Portulacaria afra
Vitex trifolia 'Purpurea'

APPENDIX B

PLANTS FOR THE BEACH GARDEN

The following recommendations for the beach garden are divided into two lists: List I includes those species thriving in the face of strong, onshore winds; List II includes those requiring protection by other plants or planted in the lee of the beach house. The amount of protection will vary with the species and the air quality of the site. In a shore situation with a reef far out, minimal protection will be required. On the other hand, a site with a close, strong shorebreak will require maximum protective plantings.

A basic assumption is made: species in the following genera are not listed below, as we feel that all are capable of tolerating List II conditions and, probably, List I conditions as well: *Aloe, Agave, Echeveria, Euphorbia, Crassula, Kalanchoe,* and *Sansevieria.*

Native Hawaiian and Heritage species are followed by (H).

LIST I Wind Tolerant

GROUND COVERS AND SMALL SPECIES TO 2 FEET

Carpobrotus deliciosus
C. edulis
Gaillardia x *grandiflora*
Heliotropium anomalum var. *argenteum* (H)
Ipomoea carica (H)
I. pes-caprae subsp. *brasiliensis* (H)
Jaquemontia ovalifolia subsp. *sandwicensis* (H)
Melanthera integrifolia (H)
M. lobata (H)
Plectranthus parviflorus (H)
Portulaca lutea (H)
P. molokiniensis (H)
Pseudognaphalium sandwicensium (H)
Scaevola coriacea (H)
Sesuvium portulacastrum (H)
Sida fallax (H)
Stapelia gigantea
Tetragonia tetragonioides
Zinnia palmeri

SMALL SHRUBS: 2-6 FEET

Adenium swazicum
Agapanthus praecox subsp. *orientalis*
Bryophyllum gastonis-bonnieri
Capparis sandwichiana (H)
Carissa macrocarpa 'Prostrata'
Colubrina asiatica (H)
Crinum asiaticum
Gossypium tomentosum (H)
Jatropha podagrica
Lantana camara (modern hybrids and cultivars)
Lycium sandwicense (H)
Myoporum sandwicense (creeping form) (H)
Pedilanthus macrocarpus
Pittosporum tobira 'Wheeler's Dwarf'
Santalum ellipticum (H)
Sesbania tomentosa (H)
Sida fallax (H)
Vitex rotundifolia (H)
Wikstroemia uva-ursi (H)

MEDIUM SHRUBS: 6-10 FEET

Adenium arabicum
A. multiflorum
A. obesum
A. obesum subsp. *socotranum*
A. somalense
Chrysobalanus icaco
Clerodendrum inerme
Pedilanthus tithymaloides subsp. *tithymaloides*
Pittosporum tobira 'Variegata'
Scaevola taccada (H)
Synadenium grantii

LARGE SHRUBS: OVER 10 FEET

Bontia daphnoides
Calotropis gigantea
C. procera
Carissa macrocarpa
Ficus microcarpa var. *crassifolia*
Pittosporum tobira
Portulacaria afra
Tamarix chinensis
Vitex trifolia 'Purpurea'
Yucca aloifolia

SMALL TREES: 15-30 FEET

Clusia rosea
Morinda citrifolia (H)
Myoporum sandwicense (H)
Plumeria obtusa (Bahamas)
Scaevola taccada (tree form)
Tournefortia argentea

MEDIUM TREES: 30-50 FEET

Coccoloba uvifera
Conocarpus erectus
Noronhia emarginata
Terminalia catappa

LARGE TREES: OVER 50 FEET

Araucaria columnaris
Calophyllum inophyllum

LIST II Some Wind Protection Recommended

GROUND COVERS AND SMALL SPECIES TO 2 FEET

Aeollanthus repens
Bryophyllum fedtschenkoi
Callisia repens
Cyanotis somaliensis
Gasteria sp.
Graptopetalum paraguayense
Haworthia aristata
H. attenuata
H. coarctata
H. fasciata
H. greenii
H. limifolia
Pachyphytum compactum
Plectranthus amboinicus
P. prostratus
Portulaca grandiflora
Sedum adolphi
S. lucidum
S. x *rubrotinctum*
Senecio mandraliscae
S. vitalis
Stapelia gigantea
Tradescantia pallida 'Purpurea'
T. sillimontana
T. spathacea
Verbena x *hybrida*

SMALL SHRUBS: 2–6 FEET

Achyranthes splendens (H)
A. splendens var. *rotundata* (H)
Argemone glauca (H)
A. mexicana
Asparagus densiflorus 'Meyers'
A. densiflorus 'Sprengeri'
Bougainvillea 'Temple Fire'
Calandrinia grandiflora
Catharanthus roseus
Dasylirion wheeleri
Euryops chrysanthemoides x *E. speciosissimus*
E. pectinatus subsp. pectinatus
Hibiscus ovalifolius
Jasminum sambac
Juniperus procumbens
Lantana montevidensis
Nototrichium divaricatum (H)
Plumbago zeylanica (H)
Rosmarinus officinalis
Senecio cineraria
S. crassissimus
Serissa japonica
Zamioculcas zamiifolia

MEDIUM SHRUBS: 6–10 FEET

Brighamia insignis (H)
B. rockii (H)
Chadsia grevei
Coprosma repens 'Picturata'
Ficus deltoidea
Juniperus chinensis 'Pfitzeriana'
Nerium oleander 'Petite Pink' et al.
Nototrichium sandwicense (H)
Osteomeles anthyllidifolia (H)
Plumbago auriculata
Pseuderanthemum carruthersi var. *atropurpureum*
P. carruthersi var. *reticulatum*
P. carruthersi var. *variegatum*
Rhaphiolepis indica
R. umbellata var. *integerrima*
Sanchezia speciosa
Strelitzia reginae
S. reginae var. *juncea*
Tecoma capensis

LARGE SHRUBS: OVER 10 FEET

Acalypha wilkesiana
A. wilkesiana 'Picotee'
Codiaeum variegatum
Furcraea foetida 'Mediopicta'
Hibiscus rosa-sinensis
Lysiloma thornberi
Murraya paniculata
Nerium oleander
Polyscias guilfoylei
Thunbergia erecta

SMALL TREES: 15–30 FEET

Anacardium occidentale
Annona muricata
Callistemon citrinus
C. viminalis
Chloroleucon tortum
Cordia lutea
C. sebestena
Delonix decaryi
Dracaena marginata
D. marginata 'Grandifolia'
Gossypium barbadense
Guaiacum officinale
G. sanctum
Ilex vomitoria
Jatropha multifida
Melaleuca bracteata 'Revolution Gold'
Metrosideros polymorpha (H)
Munroidendron racemosum (H)
Plumeria caracasana
P. obtusa var. *sericifolia*
Psydrax odorata (H)
Thevetia peruviana
T. thevetioides

MEDIUM TREES: 30–50 FEET

Ceratonia siliqua
Chorisia speciosa
Cordia subcordata
Cupaniopsis anacardioides
Delonix regia
Elaeodendron orientale
Erythrina sandwicensis
Ficus lyrata
Lagunaria patersonii
Lonchocarpus punctatus
L. violaceus
Lysiloma latisiliquum
Manilkara zapota
Olea europaea subsp. *cuspidata*
Piscidia piscipula
Plumeria obtusa
Podocarpus elatus
P. macrophyllus
Prosopis pallida
Thespesia populnea (H)

LARGE TREES: OVER 50 FEET

Andira inermis
Azadirachta indica
Barnebydendron riedelii
Brosimum alicastrum
Bucida buceras
Bursera simaruba
Cassia x *nealiae*
Enterolobium cyclocarpum
Erythrina variegata var. *orientalis*
Ficus benghalensis
F. benjamina
F. macrophylla
Tabebuia heterophylla
Tamarindus indica

VINES

Bougainvillea spectabilis
Cissus rotundifolia
Petrea volubilis
Podranea ricasoliana
Pseudogynoxys chenopodioides
Trachelospermum jasminoides
Tristellateia australasiae

BROMELIADS

Aechmea blanchetiana
Alcantara imperialis
Ananas comosus var. *variegata*

CYCADS

Cycas circinalis
C. revoluta
Encephalartos lehmannii
E. trispinosus
Zamia furfuracea
Z. integrifolia

APPENDIX C

NATIVE HAWAIIAN AND "HERITAGE" PLANTS FOR THE XERISCAPE

GROUND COVERS AND SMALL SPECIES TO 2 FEET

Bidens hillebrandiana
B. mauiensis
Heliotropium anomalum var. *argenteum*
Ipomoea cairica
I. pes-caprae subsp. *brasiliensis*
Jaquemontia ovalifolia subsp. *sandwicensis*
Melanthera integrifolia
M. lobata
Plectranthus parviflorus
Portulaca lutea
P. molokiniensis
Pseudognaphalium sandwicensium
Scaevola coriacea
Sida fallax
Tetramalopium filiforme

SMALL SHRUBS: 2–6 FEET

Abutilon menziesii
Achyranthes splendens
A. splendens var. *rotundata*
Argemone glauca
Atemisia australis
A. mauiensis
Bidens cosmoides
Colubrina asiatica
Gossypium tomentosum
Melanthera lavarum
Myoporum sandwicense (creeping form)
Nototrichium divaricatum
Plumbago zeylanica
Santalum ellipticum
Sesbania tomentosa
Sida fallax
Vitex rotundifolia
Wikstroemia uva-ursi

MEDIUM SHRUBS: 6–10 FEET

Bidens menziesii
B. menziesii subsp. *filiformis*
B. micrantha subsp. *ctenophylla*
Brighamia insignis
B. rockii
Hibiscus brackenridgei subsp. *brackenridgei*
H. brackenridgei subsp. *mokuleianus*
Nototrichium sandwicense
Osteomeles anthyllidifolia
Scaevola taccada

LARGE SHRUBS: OVER 10 FEET

Dodonaea viscosa
Wilkesia gymnoxiphium

SMALL TREES: 15–30 FEET

Gardenia brighamii
Kokia drynarioides
Metrosideros polymorpha
Morinda citrifolia
Munroidendron racemosum
Myoporum sandwicensis
Pittosporum confertiflorum
P. flocculosum
Pleomele hawaiiensis
Psydrax odorata
Sophora chrysophylla

MEDIUM TREES: 30–50 FEET

Colubrina oppositifolia
Cordia subcordata
Erythrina sandwicensis
Euphorbia haeleeleana
Sapindus oahuensis
S. saponaria
Thespesia populnea

LARGE TREES: OVER 50 FEET

Calophyllum inophyllum
Reynoldsia sandwicensis

VINES

Bonamia menziesii
Canavalia galeata
C. hawaiiensis
C. pubescens

APPENDIX D

PLANTS FOR THE NIGHTSCAPE

Lighting the garden for evening use can be a challenge and a delightful success. Too frequently, simple floodlights are aimed at the plantings with the result that plants become visually "flattened" and the result dull. Uplighting, on the other hand, if not overdone, results in lighting the tree or shrub almost from within, creating a glowing result and providing visual depth, foliage, and bark color and dramatizing any special structural patterns of branching or translucent foliage (e.g., *Scaevola taccada, Tournefortea argentea,* et al.). Lighting is especially effective on unusual bark color or pattern (e.g., *Chloroleucon tortum, Pimenta racemosa* and *P.dioica, Acacia xanthophloea, Acacia seyal* var. *seyal,* et al.). In certain situations a combination of uplighting and minifloods will provide the best results. Minifloods are especially effective in casting a trunk/branching silhouette against a light-colored wall. In no case should the source of light be seen. And less lighting is better than too much lighting. The small ground covers can be downlighted by short, hooded fixtures a foot or so tall to line a walkway or a change in the grade or direction of a path. There are many excellent designs available, some especially for beach gardens where corrosion is a problem. Low voltage systems are recommended. There are many to choose from. In any event, work with your landscape architect to design a night garden. Or work with the supplier if you embark on a do-it-yourself lighting system. Use a licensed electrician for advice and final connections.

In the following list, species name followed by (H) indicates a Hawaiian or Hawaiian introduced species.

GROUND COVERS AND SMALL SPECIES TO 2 FEET

Aeollanthus repens
Graptopetalum paraguayense
Heliotropium anomalum var. *argenteum* (H)
Kalanchoe farinacea
K. pumila
K. tomentosa
Pseudognaphalium sandwicensium (H)
Senecio mandraliscae
S. vitalis
Tradescantia pallida 'Purpurea'

SMALL SHRUBS: 2–6 FEET

Achyranthes splendens (H)
Argemone glauca (H)
Artemesia australis (H)
A. mauiensis (H)
Capparis sandwichiana (H)
Crinum asiaticum
Kalanchoe orygalis
Nototrichium divaricatum (H)
Olea europaea 'Little Ollie'
Russelia equisetiformis
Santalum ellipticum (H)
Senecio cineraria
S. crassissimus
Sesbania tomentosa (H)
Wikstroemia uva-ursi (H)

MEDIUM SHRUBS: 6–10 FEET

Agave americana
A. americana var. *marginata* (et al.)
A. attenuata
Brighamia insignis (H)
B. rockii (H)
Furcraea foetida 'Mediopicta'
Kalanchoe beharensis
Leucophyllum frutescens
Nototrichium sandwicense (H)
Pseuderanthemum carruthersii var. *atropurpureum*
P. carruthersii var. *reticulatum*
P. carruthersii var. *variegatum*
Scaevola taccada (H)

LARGE SHRUBS: OVER 10 FEET

Acalypha wilkesiana
Breynia disticha 'Roseopicta'
Calotropis gigantea
C. procera
Codiaeum variegatum
Euphorbia lactea var. *variegata* 'White Ghost'
Kalanchoe hildebrandtii
Tamarix chinensis
Vitex trifolia 'Purpurea'

SMALL TREES: 15–30 FEET

Acacia seyal var. *seyal*
A. seyal var. *fistula*
Chloroleucon tortum
Dracaena draco
D. marginata (foliage color forms)
Euphorbia cotinifolia
Guaiacum officinale
G. guatemalense
G. sanctum
Ipomoea arborescens
Meleleuca bracteata 'Revolution Gold'
Scaevola taccata (tree form)
Sesbania grandiflora
Thevetia peruviana
T. thevetioides
Tournefortia argentea

MEDIUM TREES: 30–50 FEET

Caesalpinia ferrea
Conocarpus erectus
Eucalyptus bigalerita
Lagunaria patersonii
Melaleuca argentea
Moringa drouhardii
M. stenopetala
Olea europaea subsp. *europaea*
Pithecellobium dulce 'Variegata'
Pimenta dioica
P. racemosa
Prosopis pallida
Pseudobombax ellipticum

LARGE TREES: OVER 50 FEET

Acacia xanthophloea
Adansona digitata
Bursera simaruba
Enterolobium cyclocarpum
Eucalyptus citriodora
Ficus benghalensis
F. sycomorus subsp. *gnaphalocarpa*
Reynoldsia sandwicensis (H)

VINES

Argyreia nervosa
Bonamia menziesii (E)
Bougainvillea spectabilis
Congea tomentosa

BROMELIADS

Aechmea blanchetiana
Alcanteara imperialis
Ananas comosus var. *variegata*
Tillandsia rothii
T. tectorum
T. xerographica

CYCADS

Encephalartos eugene-maraisii
E. horridus
E. lehmannii
Macrozamia macdonnellii
M. moorei

APPENDIX E

COLOR LISTS FOR THE XERISCAPE

Following are lists of colors (flowers, foliage, bark, fruit) that will assist the xeriscaper in designing his landscape. Colors are generic; that is, yellow may be light yellow, lemon yellow, or gold. Look closely at the photos in the text. Some plants are listed more than once, as they may have flowers of a number of colors. Plumeria is a good example. All represent flower color unless otherwise noted.

YELLOW

GROUND COVERS AND SMALL SPECIES TO 2 FEET

ZONE 2

Tagetes patula

ZONE 3

Coreopsis grandiflora
C. lanceolata hybrids
Kalanchoe blossfeldiana
Portulaca grandiflora

ZONE 4

Aloe aculeata
A. hildebrandtiii
A. sinkatana
Bulbine frutescens
Bulbinella floribunda
Carpobrotus edulis
Euphorbia decaryi
Melanthera integrifolia (H)
M. lobata (H)
Portulaca lutea
P. molokiniensis (H)
Pseudognaphalium sandwicensium (H)
Senecio vitalis
Sida fallax (H)
Zinnia palmeri

SMALL SHRUBS: 2–6 FEET

ZONE 2

Allamanda cathartica 'Thai Dwarf'
A. schotti
Tagetes erecta

ZONE 3

Euryops chrysanthemoides x *E. speciosissimus*
Hibiscus ovalifolius
Justicia brandegeana
Russelia equisetiformis
Sida fallax (H)
Zinnia violacea

ZONE 4

Agave bracteosa
A. gypsophila
A. vivipara 'Marginata'
Aloe broomii
A. saponaria
A. vera
Argemone mexicana
Bryophyllum gastronis-bonnieri
Euphorbia milli
Gossypium tomentosum (H)
Lantana camara hybrids
Melanthera lavarum (H)
Sansevieria trifasciata foliage
Senecio crassissimus
Sesbania tomentosa (H)

MEDIUM SHRUBS: 6–10 FEET

ZONE 2

Bidens cosmoides (H)
Galphimia gracilis
Pseuderanthemum carruthersi var. *reticulatum* foliage

ZONE 3

Allamanda laevis
Bidens menziesii (H)
Hibiscus brackenridgei subsp. *brackenridgei* (H)

ZONE 4

Agave americana foliage
Aloe arborescens
Cordeauxia edulis
Furcraea foetida 'Mediopicta' foliage

LARGE SHRUBS: OVER 10 FEET

ZONE 2

Acalypha wilkesiana foliage
Bixa orellana fruit
Hibiscus rosa-sinensis hybrids

ZONE 3

Codiaeum variegatum foliage
Duranta erecta 'Gold' foliage/fruit
Holmskioldia sanguinea
Senna polyphylla

ZONE 4

Aloe castanea
Nerium oleander

SMALL TREES: 15–30 FEET

ZONE 2

Melaleuca bracteata 'Revolution Gold'

ZONE 3

Bauhinia tomentosa
Diphysa americana
Metrosideros polymorpha (H) [Note: although eventually

reaching as much as 100 feet in height, its slow growth suggests treating it as small tree.]

Parmentiera aculeata
Pterocarpus rohrii
Tabebuia aurea
Uncarina grandidieri

ZONE 4

Caesalpinia pulcherrima
Cordia lutea
Gossypium barbadense
Senna surratensis
Thevetia peruviana
T. thevetioides

MEDIUM TREES: 30-50 FEET

ZONE 2

Deplanchea tetraphylla
Peltophorum pterocarpum

ZONE 3

Caesalpinia ferrea
Cassia fistula
C. sieberiana
Delonix regia var. *flava*
Plumeria rubra hybrids
Thespesia populnea

LARGE TREES: OVER 50 FEET

ZONE 2

Pterocarpus indicus
Tabebuia ochracea subsp. *neochrysantha*

ZONE 3

Acacia xanthophloea bark
Byrsonima crassifolia flowers and fruit
Cassia x *nealiae* 'Lunalilo Yellow', 'Queen's Hospital White', 'Nii Gold'
Cassia x *nealiae* 'Wilhelmina Tenney': bicolor: petals yellow on inside, cerise on outside
Cochlospermum vitifolium
Platymiscium pinnatum
Pterocarpus marsupium
Tabebuia chrysea
T. donnell-smithii

ZONE 4

Reynoldsia sandwicensis (H)
Tamarindus indicus
Tipuana tipu

VINES

ZONE 2

Allamanda cathartica
Stigmaphyllon ciliatum

ZONE 3

Bougainvillea hybrids
Tristellateia australasiae

BROMELIADS

ZONE 3

Aechmea blanchetiana foliage

ORANGE

GROUND COVERS AND SMALL SPECIES TO 2 FEET

ZONE 2

Tagetes patula

ZONE 3

Coreopsis grandiflora
C. lanceolata
Gaillardia x *grandiflora*
Gazania rigens
Kalanchoe blossfeldiana

ZONE 4

Aloe aculeata
A. dorotheae
A. hildebrandtii
A. x *nobilis*
A. sinkatana
Bulbine frutescens 'Hallmark'
Echeveria carnicolor

SMALL SHRUBS: 2-6 FEET

ZONE 2

Pelargonium x *hortorum*
Tagetes erecta

ZONE 3

Aloe camperi
Anisacanthus thurberi
Sida fallax (H)
Zinnia violacea

ZONE 4

Aloe maculata
A. striatula
A. vera var. *chinensis*
Lantana camara hybrids

MEDIUM SHRUBS: 6-10 FEET

ZONE 3

Chadsia grevei
Rondeletia odorata
Strelitzia reginae
Tecoma capensis

ZONE 4

Aloe ferox

LARGE SHRUBS: OVER 10 FEET

ZONE 2

Hibiscus rosa-sinensis hybrids

ZONE 3

Duranta erecta

ZONE 4

Aloe africana
A. excelsa
A. marlothii
Nerium oleander

SMALL TREES: 15-30 FEET

ZONE 2

Eriobotrya japonica

ZONE 3

Cordia sebestena
Dracaena draco fruit
Guaiacum officinale fruit
G. sanctum fruit
Metrosideros polymorpha (H)

ZONE 4

Jacquinia pungens

MEDIUM TREES: 30-50 FEET

ZONE 3

Butea frondosa
Cordia subcordata
Plumeria rubra hybrids
Terminalia catappa

ZONE 4

Colvillea racemosa

LARGE TREES: OVER 50 FEET

ZONE 3

Mangifera indica

VINES

ZONE 3

Bougainvillea hybrids
Pseudogynoxys chenopodioides

BROMELIADS

ZONE 3

Aechmea blanchetiana foliage

RED

GROUND COVERS AND SMALL SPECIES TO 2 FEET

ZONE 2

Tagetes patula

ZONE 3

Gaillardia x *grandiflora*
Kalanchoe blossfeldiana
Portulaca grandiflora
Verbena x *hybrida*

ZONE 4

Aloe ammophila
A. brevifolia
A. deltoideodonta
A. greatheadii
A. hildebrandtii
A. rivae
A. sinkatana
A. squarrosa
A. zebrina
Crassula capitella 'Camp Fire' foliage
Echeveria agavoides var. *prolifera* foliage
Euphorbia decaryi
Kalanchoe farinacea
K. luciae foliage
Sedum x *rubrotinctum* foliage

SMALL SHRUBS: 2–6 FEET

ZONE 2

Callistemon viminalis 'Little John'
Cuphea hybrids
Pelargonium x *hortorum*

ZONE 3

Aloe berhana
A. cameronii
A. dawei
Bougainvillea 'Temple Fire'
Carissa macrocarpa 'Prostrata' fruit
Catharanthus roseus
Crassula perfoliata var. *falcata*
Justicia brandegeana
Malvaviscus arboreus var. *drummondii*
Russelia equisetiformis
Wikstroemia uva-ursi fruit
Zinnia violacea

ZONE 4

Aloe chabaudii
A. distans
A. framesii
A. globuligemma
A. mitriformis
A. petricola
A. x *salm-dychiana*
A. saponaria
A. suarezensis
Crassula ovata foliage
Euphorbia milii
Lantana camara hybrids
Pedilanthus macrocarpus
Sesbania tomentosa (H)

MEDIUM SHRUBS: 6–10 FEET5

ZONE 2

Sanchezia speciosa

ZONE 3

Chrysobalanus icaco foliage
Punica granatum 'Nana' flowers and fruit
Rondoletia odorata

ZONE 4

Adenium multiflorum
A. obesum
A. somalense
Aloe arborescens
A. ferox
A. vaombe
Synadenium grantii foliage

LARGE SHRUBS: OVER 10 FEET

ZONE 2

Acalypha wilkesiana foliage
Calliandra haematocephala
Hibiscus rosa-sinensis hybrids
Hibiscus schizopetalus

ZONE 3

Carissa macrocarpa fruit
Codiaeum variegatum foliage
Holmskioldia sanguinea

ZONE 4

Aloe africana
A. excelsa
A. kedongensis
A. volkensii
Dodonaea viscosa (H) fruit
Nerium oleander

SMALL TREES: 15–30 FEET

ZONE 2

Callistemon citrinus
C. rigidus
C. viminalis
Euphorbia cotinifolia foliage
Ilex vomitoria fruit

ZONE 3

Cassia roxburghii
Dracaena marginata
Erythrina latissima
E. madagascariensis
Euphorbia punicea
Kokia drynarioides (H)
Majidea zanquebarica fruit
Malpighia emarginata fruit
Metrosideros polymorpha (H)

ZONE 4

Acacia seyal var. *seyal* bark
Anacardium occidentale foliage
Caesalpinia pulcherrima
Euphorbia tirucalli 'Sticks on Fire' bark
Fouquieria macdougalii
Punica granatum flowers and fruit

MEDIUM TREES: 30–50 FEET

ZONE 3

Chorisia speciosa 'Los Angeles Beautiful'
Delonix regia
Erythrina abyssinica
E. lysistemon
Ficus ingens foliage
Plumeria rubra hybrids
Schotia brachypetala flowers and foliage
Terminalia catappa foliage

LARGE TREES: OVER 50 FEET

ZONE 2

Hildebrandia barteri

ZONE 3

Barnebydendron riedlii
Bombax ceiba
Erythrina sacleuxii
Sterculia foetida fruit

PINK

GROUND COVERS AND SMALL SPECIES TO 2 FEET

ZONE 2:

Ipomoea pes-caprae subsp. *brasiliensis*

ZONE 3:

Aloe haworthioides x *descoingsii*

ZONE 4:

Aloe juvenna
A. striata
Bryophyllum fedtschenkoi
Carpobrotus deliciosus
Echeveria 'Morning Light' flowers and foliage
Gasteria carinata
Kalanchoe marnieriana

SMALL SHRUBS: 2–6 FEET

ZONE 2:

Pelargonium x *hortorum*
Plumeria 'Petite Pink'

ZONE 3:

Ipomoea carnea subsp. *fistulosa*
Malvaviscus arboreus var. *drummondii*
Russelia equisetiformis
Zinnia violacea

ZONE 4 :

Adenium swazicum
Euphorbia milii
Lantana camara hybrids

MEDIUM SHRUBS: 6–10 FEET

ZONE 3:

Bauhinia grandidieri
Leucophyllum frutescens
Nerium olender 'Petite Pink'
Rhaphiolepis indica

ZONE 4:

Adenium arabicum
A. obesum
A. somalense
Punica protopunica flowers and fruit

LARGE SHRUBS: OVER 10 FEET

ZONE 2:

Hibiscus rosa-sinensis hybrids
H. schizopetalus hybrids

ZONE 3:

Codaeium variegatum foliage

ZONE 4:

Nerium oleander

SMALL TREES: 15–30 FEET

ZONE 3:

Lagerstroemia indica
Malpighia emarginata
Metrosideros polymorpha (H)
Sesbania grandiflora
Tabebuia berteroi

ZONE 4:

Caesalpinia pulcherrima
Lysiphyllum cunninghamii
Pereskia grandifolia

MEDIUM TREES: 30–50 FEET

ZONE 2:

Bauhinia x *blakeana*
Cassia grandis
Eucalyptus bigalerita

ZONE 3:

Chorisia speciosa 'Majestic Beauty'
Erythrina dominguezii
Lagunaria patersonii
Piscidia piscipula
Plumeria rubra hybrids
Pseudbombax ellipticum

LARGE TREES: OVER 50 FEET

ZONE 3:

Samanea saman
Tabebuia heterophylla
T. impetiginosa
T. rosea

LAVENDER

GROUND COVERS AND SMALL SPECIES TO 2 FEET

ZONE 4:

Cyanotis somaliensis
Graptopetalum paraguayense
Kalanchoe pumila flowers and foliage
K. synsepala
Plectranthus prostratus
Sesuvium portulacastrum (H)
Tetramolopium filiforme (H)

SMALL SHRUBS: 2–6 FEET

ZONE 2:

Pelargonium x *hortorum*

ZONE 3:

Zinnia violacea

ZONE 4:

Adenium swazicum
Lantana montevidensis

MEDIUM SHRUBS: 6–10 FEET

ZONE 3:

Leucophyllum frutescens

LARGE SHRUBS: OVER 10 FEET

ZONE 2:

Hibiscus rosa-sinensis hybrids

ZONE 3:

Duranta erecta 'Gold'
Grewia occidentalis

ZONE 4:

Calotropis gigantea

SMALL TREES; 15–30 FEET

ZONE 3:

Lagerstroemia indica

MEDIUM TREES: 30–50 FEET

ZONE 3:

Lonchocarpus punctatus
L. violaceus

LARGE TREES: OVER 50 FEET

ZONE 3:

Andira inermis

BLUE

GROUND COVERS AND SMALL SPECIES TO 2 FEET

ZONE 3:

Verbena x *hybrida*

ZONE 4:

Aeollanthus repens
Jaquemontia ovalifolia subsp. *sandwicensis* (H)
Senecio mandraliscae bark
S. vitalis foliage

SMALL SHRUBS: 2–6 FEET

ZONE 2:

Agapanthus praecox subsp. *orientalis*
Callistemon viminalis 'Little John' foliage
Evolvulus glomeratus subsp. *grandiflorus*

ZONE 3:

Rosmarinus officinalis

ZONE 4:

Argemone glauca foliage
Vitex rotundifolia

MEDIUM SHRUBS: 6–10 FEET

ZONE 3:

Petrea sp.
Plumbago auriculata
Strelitzia reginae

ZONE 4:

Agave attenuata foliage
Euporbia coerulescens bark

LARGE SHRUBS: OVER 10 FEET

ZONE 3:

Duranta erecta 'Geisha Girl'

SMALL TREES: 15–30 FEET

ZONE 3:

Bolusanthus speciosus
Dracaena draco foliage
Guaiacum officinale
G. sanctum

ZONE 4:

Guaiacum guatemalense

MEDIUM TREES: 30–50 FEET

ZONE 3:

Jacaranda mimosifolia

VINES

ZONE 3:

Petrea volubilis

CYCADS

The following small group of cycads from South Africa and Australia carries bluish gray foliage and may be useful in planning a colorful xeriscape.

SOUTH AFRICAN

ZONE 3:

Encephalartos horridus
E. lehmannii
E. trispinosus

ZONE 4:

Encephalartos eugene-maraisii

AUSTRALIAN

ZONE 3:

Macrozamia macdonnellii
M. moorei

PURPLE

GROUND COVERS AND SMALL SPECIES TO 2 FEET

ZONE 3:

Carpobrotus deliciosus
Kalanchoe blossfeldiana
Tradescantia pallida 'Purpurea' foliage
T. zebrina 'Discolor'
T. zebrina foliage
Verbena x *hybrida*

ZONE 4:

Echeveria carnicolor foliage
Pachyphytum compactum foliage

SMALL SHRUBS: 2–6 FEET

ZONE 2:

Agapanthus praecox subsp. *orientalis*
Crinum asiaticum foliage

ZONE 3:

Calandrinia grandiflora
Catharanthus roseus

ZONE 4:

Senecio crassissimus foliage

MEDIUM SHRUBS: 6–10 FEET

ZONE 2:

Pseuderanthemum carruthersi var. *atropupureum* foliage

ZONE 3:

Leucophyllum frutescens

LARGE SHRUBS: OVER 10 FEET

ZONE 2:

Thunbergia erecta

ZONE 3:

Duranta erecta 'Goldedge'

ZONE 4:

Calotropis procera
Vitex trifolia 'Purpurea' foliage

SMALL TREES: 15–30 FEET

MEDIUM TREES: 30–50 FEET

ZONE 2:

Bauhinia x *blakeana*
B. variegata

ZONE 3:

Lonchocarpus sericeus

ZONE 4:

Kigelia africana

LARGE TREES: OVER 50 FEET

ZONE 3:

Lonchocarpus domingensis

VINES

ZONE 2:

Allamanda blanchetii
Saritaea magnifica

ZONE 3:

Bougainvillea hybrids

ZONE 4:

Canavalia galeata (H)

BROMELIADS

ZONE 3:

Alcantarea imperialis foliage

For plants with gray silver foliage, refer to the list of those suggested for the nightscape in Appendix D.

APPENDIX F

EVAPOTRANSPIRATION

Evapotranspiration (ET) is a term describing the estimated total amount of water lost to the atmosphere through the combined processes of evaporation from the plant surfaces and soil in the plant's root zone and water lost as vapor or gas through stomata in the plant's leaves. This amount of lost water usually needs to be replaced, all or in part, for normal growth. The amount of lost water is typically expressed as a depth in inches (or millimeters) over a specified period of time. Many states and water agencies generate and provide ET data, which irrigation managers and other landscape professionals can use to help them more accurately schedule irrigation frequencies and run times to apply only the amount of water needed.

Many factors affect ET, including weather (temperature, relative humidity, solar radiation, wind, etc.); soil conditions (texture, structure, etc.); and the plant species. Because so many factors affect ET, it is impossible to have one set of ET data that is reliable for all plant species at all sites under all conditions. Thus, scientists have developed the concept of reference ET (ETo), which is the amount of ET from a standardized grass surface (tall fescue) at a given site. ET amounts for individual plant species are a percentage of the ETo. This percentage is called a crop factor or crop coefficient (Kc) and is used to calculate the ET for a particular plant species (ETc). For example, if the Kc for a particular tree species was 0.5 and the ETo was 0.5 inches for the day, the ETc for that tree would be 0.25 inches (0.5 x 0.5 = 0.25) for the day, which is the amount of water that would need to be applied over the tree's effective root zone (from the trunk out to at least the canopy drip line) for normal growth.

Because irrigation on a daily basis is horticulturally unsound, irrigation managers typically wait until daily ETc accumulates to about 50 percent of available water held in the upper foot of soil before replacing that amount of water in an irrigation event. For example, clay soils have about 2 inches of available water per foot of soil profile. Thus, our tree growing on a clay soil would be irrigated when daily ETc accumulates to 50 percent of this amount, or 1 inch (0.5 x 2.0 = 1.0). If daily ETc continued to accumulate at 0.25 inches per day as in our example above, our tree would be irrigated after four days (1.0 ÷ 0.25 = 4), and then 1 inch of water would be applied at that time.

Many agencies provide ETc data for numerous sites on a daily or real-time basis through online access of computer-driven weather stations. Historical ETo data, based on numerous years of records, are also sometimes available. Unfortunately, specific Kc values for many landscape plant species are lacking. However, general Kc values for most non–rain forest and nondesert plants range from 0.5 to 0.75. Desert plants will have lower and rain forest plants higher Kc values. Keep in mind that ET data are just one of several tools that irrigation managers have at their disposal to help them schedule irrigations accurately.

APPENDIX G

HAWAII-PACIFIC WEED RISK ASSESSMENT PROJECT

This is a joint project of the University of Hawai'i and Kaulunani Urban and Community Forestry, a program of the State of Hawaii Department of Land and Natural Resources, Division of Forestry and Wildlife and the United States Department of Agriculture Forest Service. The intent of the project is to identify plants that pose a high weed risk in Hawai'i. There are four basic designations:

- L and L (Hawai'i) indicates that species are not currently recognized as invasive. These designations are not included in this volume's species listings. (Hawai'i) indicates that the listed species is "documented to cause significant ecological or economic harm to Hawai'i." These species are not recommended by the authors for planting even though they are commonly used landscape species. They include *Psidium cattleianum* (strawberry guava), *Psidium guajava* (common guava), *Schefflera actinophylla* (octopus tree), and *Schinus terebinthifolius* (Christmas berry tree).

- (HPWRA) indicates that the listed species is "likely to be invasive in Hawai'i." This is a prediction but lacks in-the-field evaluation and, therefore, it is not yet documented. These species include *Bauhinia monandra* (pink bauhinia), *Bauhinia purpurea* (purple orchid tree), *Cinnamomum verum* (cinnamon tree), *Clerodendrum quadriloculare* (starburst), *Coffea arabica* (Arabian coffee), *Duranta erecta* (golden dewdrop), *Eugenia uniflora* (Surinam cherry), *Ligustrum sinense* (Chinese privet), *Morinda citrifolia* (*noni*), *Tecoma stans* (yellow elder), and *Thevetia peruviana* (be-still tree). They are recommended for planting but with the request that if any sign of invasiveness is seen, both parent plant and all seedlings be destroyed.

- Several species are in the project's "Evaluate" list and have been included in the text. The "Evaluate" list reflects lack of important information or the difficulty of assessing them using the project system. They are, however, valuable landscape species and include *Clusia rosea* (autograph tree), *Eriobotrya japonica* (loquat), *Ligustrum japonicum* (Japanese privet), *Murraya paniculata* (mock orange tree), *Pittosporum tobira* (Japanese pittosorum), *Pittosporum viridiflorum* (Cape pittosporum) and *Senna surratensis* (scrambled eggs, *kolomona*). The text is also in compliance with the constraints published by the Landscape Industry Council of Hawaii.

8
8

SUGGESTED READINGS

Adams, C. D. *Flowering Plants of Jamaica*. Mona, Jamaica: University of the West Indies (pub. Robert MacLeho), 1972.

Anderson, Miles. *The Ultimate Book of Cacti and Succulents*. New York: Lorenz Books (imprint of Anness Publishing), 1998.

Bailey, Liberty Hyde, Ethel Zoe Bailey, et al. *Hortus Third: A Concise Dictionary of Plants Cultivated in the United States and Canada*. New York: Macmillan, 1976.

Barwick, Margaret. *Tropical and Subtropical Trees: An Encyclopedia*. Portland, OR: Timber Press, 2004.

Beentje, H. J. *Kenya Trees, Shrubs and Lianas*. Nairobi, Kenya: National Museums of Kenya, 1994.

Benthall, A. P. *The Trees of Calcutta*. Calcutta: Thacker, Spink, 1933.

Blatter, Ethelbert, and Walter Samuel Millard. *Some Beautiful Indian Trees*. 2nd ed. Ed. W. T. Stearn. Bombay, India: Bombay Natural History Society, 1954.

Brandis, Dietrich. *Indian Trees*. Dehradun, India: Bishen Singh Mahendra Pal Singh, 1971.

Breitung, August J. *The Agaves (The Cactus and Succulent Journal 1968 Yearbook)*. Reprint. Reseda, CA: Abbey Garden Press, 1968.

Britton, N. L., and J. N. Rose. *The Cactaceae*. 4 vols. New York: Carnegie Institution, 1937.

Brock, John. *Native Plants of Northern Australia*. Victoria: Reed Books, 1993.

Button, Nathaniel L., and Charles F. Millspaugh. *The Bahama Flora*. New York: Pub. by the authors, 1920.

Cactus and Succulent Journal. Claremont, CA: Cactus and Succulent Society of America, first published 1929 to date.

Carter, S., L. E. Newton, J. J. Lavanos, and C. C. Walker. *Aloes: The Definitive Guide*. London: Kew Publishing, in Association with the British Cactus and Succulent Society, 2011.

Codd, L. E. W. *Trees and Shrubs of the Kruger National Park: Botanical Survey Memoir No. 26*. Pretoria: Department of Agriculture, Division of Botany and Plant Pathology, Union of South Africa Government Printer, 1951.

Cooke, Theodore. *Flora of the Presidency of Bombay*. Vol. 1, *Botanical Survey of India by the Government of India*. Calcutta: S. N. Guha Ray, 1967.

Coyle, Jeanette, and Norman C. Roberts. *A Field Guide to the Common and Interesting Plants of Baja California*. La Jolla, CA: Natural History Books, 1975.

Dale, Ivan R., and P. J. Greenway. *Kenya Trees and Shrubs*. London: Badranam's Kenya Estates, in association with Hatchards, 1961.

Dassanayake, M. D. *A Revised Handbook to the Flora of Ceylon*. New Delhi, India: pub. for the Smithsonian Institution and the National Science Foundation by Amerind, 1983.

Denver Water Department. *Xeriscape Plant Guide*. Golden, CO: Fulcrum Publishing, 1996.

Dimmit, Mark A., and Chuck Hansen. *The Genus Adenium in Cultivation*. Tucson: Arizona-Sonora Desert Museum, 2007.

Dortort, Fred. *The Timber Press Guide to Succulent Plants of the World: A Comprehensive Reference to More than 2000 Species*. Portland, OR: Timber Press, 2011.

Du Puy, D. J., J. N. Labat, R. Rabevohitra, J. F. Villiers, J. Bosser, and J. Moat, eds. *The Leguminosae of Madagascar*. London: Royal Botanic Gardens, Kew, 2002.

Eliovson, Sima. *South African Flowers for the Garden.* Cape Town: Howard Timmins, 1957.

Ellefson, C., T. Stevens, and D. Welsh. *Xeriscape Gardening: Water Conservation for the American Landscape.* New York: Macmillan, 1992.

Flores, E. M., and W. A. Marín. *Diphysa americana.* Costa Rica: National Academy of Sciences, University of Costa Rica, School of Biology.

Folsom, Debra Brown, et al. *Dry Climate Gardening with Succulents.* New York: Huntington Botanical Gardens, Pantheon Books, Knopf Publishing Group, 1995.

Gentry, Howard Scott. *Rio Mayo Plants.* Washington, DC: Carnegie Institution of Washington Publication 527, 1942.

Haage, Walter. *Cactus and Succulents.* New York: E. P. Dutton, 1963.

Harris, Shireen. *Bulbine frutescens.* Pretoria, South Africa: Free State National Botanical Garden, June 2007.

Hillebrand, William. *Flora of the Hawaiian Islands.* New York: B. Westerman, 1888.

Hutchison, J., and J. M Dalziel, eds. *Flora of West Tropical Africa.* Vol. 3, Part 1. 2nd ed. (revised by F. N. Hepper) Millbank, London: pub. on behalf of the Governments of Nigeria, Ghana, Sierra Leone and Gambia by the Crown Agents for Overseas Governments, 1968.

Isley, Paul T., III. *Tillandsia.* Gardena, CA: Botanical Press, 1987.

Jacobson, Hermann. *A Handbook of Succulent Plants.* 3 vols. Mill Valley, CA: Strawberry Press, 1998.

Jones, David L. *Cycads of the World.* Washington, DC: Smithsonian Institution Press, 1993.

———. *Palms throughout the World.* NSW, Australia: Reed Chatswood, 1993.

Jones, Warren, and Charles Sacamano. *Plants for Dry Climates.* Tucson, AZ: Fisher Books, 2000.

Keay, R. J. W. *Trees of Nigeria.* Oxford: Clarendon Press, 1989.

Kramer, Jack. *Cacti and Other Succulents.* New York: Harry N. Abrams, 1977.

Lafon, Ron, et al., eds. *Euphorbia Journal.* Herman Schwartz. Mill Valley, CA: Strawberry Press, 1987 to date.

Lamb, Edgar. *Illustrated Reference on Cacti and Other Succulents.* 3 vols. London: Blandford Press, 1958.

Leaser, David. *Palm Trees: A Story in Photographs.* Los Angeles: Westwood Pacific, 2005.

Letty, Cythna. *Wild Flowers of the Transvaal.* Johannesburg: Department of Agriculture, Division of Botany, Pretoria Hortors, 1962.

Little, Elbert L., Jr., and Frank H. Wadsworth. *Common Trees of Puerto Rico and the Virgin Islands.* 2 vols. Washington, DC: U.S. Department of Agriculture Forest Service, 1964.

Little, Jim. *Growing Plumerias in Hawai'i.* Honolulu: Mutual Publishing, 2006.

Llamas, Kirsten A. *Tropical Flowering Plants: A Guide to Identification and Cultivation.* Portland, OR: Timber Press, 2003.

Lord, Ernest E. *Shrubs and Trees for Australian Gardens.* Melbourne and Sydney: Lothian, 1970.

Mabberley, D. J. *The Plant Book.* 2nd ed. Cambridge: Cambridge University Press, 1997.

McCann, Charles. *100 Beautiful Trees of India.* Bombay: D. B. Taraprevala Sons, 1959.

Meade, Ginny, and David L. Hensley. *Mulching for Healthier Landscape Plants.* L-3. Honolulu: University of Hawai'i at Mānoa, CTAHR, Cooperative Extension Service, October 1997. http://www.ctahr.hawaii.edu/oc/freepubs/pdf/L-3.pdf.

Menninger, Edwin A. *Seaside Plants of the World.* New York: Hearthside Press, 1964.

Milne, E., and R. M. Polhill, eds. *Flora of Tropical East Africa: (Brexiaceae B. Verdcourt).* Pub. under the authority of the Minister for Overseas Development by the Crown Agents for Overseas Governments and Administrations, 1968.

O'Gorman, Helen. *Mexican Flowering Trees and Plants.* Mexico City: Ammex Associados, 1961.

Oudhia, Pankaj. Raipur, India: Society for Parthenium Management, 2004.

Palgrave, Olive H. Coates, and Keith Coates Palgrave. *Trees of Central Africa.* Salisbury, Southern Rhodesia: National Publications Trust, 1956.

Palmer, Eve, and Nora Pitman. *Trees of South Africa.* 3 vols. Capetown: A. A. Balkema, 1972.

Pertchek, H., and Bernard Pertchek. *Flowering Trees of the Caribbean.* New York: Rinehart, 1959.

Pesman, M. Walter. *Meet Flora Mexicana.* Flagstaff, AZ: Northland Press, 1962.

Pooley, Elsa. *The Complete Field Guide to Trees of Natal, Zululand and Transkei.* Durban, South Africa: Natal Flora Publications Trust, 1994.

Rauch, Fred D., and Paul Weissich. *Plants for Tropical Landscapes.* Honolulu: University of Hawai'i Press, 2000.

———. *Small Trees for the Tropical Landscape.* Honolulu: University of Hawai'i Press, 2009.

———. *The Watersmart Garden.* Honolulu University of Hawai'i Press, 2013.

Rauh, Werner. *Succulent and Xerophytic Plants of Madagascar.* 2 vols. Mill Valley, CA: Strawberry Press, 1998

Reynolds, Gilbert W. *The Aloes of South Africa.* Johannesburg, South Africa: Trustees of the Aloe Book Fund, 1950.

———. *The Aloes of Tropical Africa and Madagascar.* Mbabane, Swaziland: Trustees of the Aloe Book Fund, 1966.

Rojo, Justo P. *Phanerogamarum Monographica Tomus.* Vol. 5: *Pterocarpus.* Germany: Verlag von J. Cramer, 1972.

Rowley, Gordon D. *Caudiciform and Pachycaul Succulents.* Mill Valley, CA: Strawberry Press, 1987.

Russell, T. C., and M. Martin. *American Trees.* London: Hermes House, 2006.

Saha, Sonali, and Henry F. Howe. "Species Composition and Fire in a Dry Deciduous Forest." *Ecology* 84, no. 12 (2003): 3118–3123.

Sajeva, Maurizo, and Mariangeela Costanza. *Succulents.* Portland, OR: Timber Press, 2001.

Sánchez de Lorenzo Cáceres, José Manuel, and Antonio

López Lillo. *Árboles en España: Manual de Identificación*. Madrid: Muni-Prensa Libros, S.A., 2004.
Schatz, George E. *Generic Tree Flora of Madagascar*. London: Royal Botanic Gardens, Kew, and St. Louis: Missouri Botanical Garden, 2001.
Shuler, C. *Low Water Use Plants for California and the Southwest*. Cambridge, MA: Fisher Books, 1993.
Standley, Paul C. *Flora of Costa Rica*. Publication 392. Chicago: Field Museum of Natural History, 1937.
——. *Flora of the Panama Canal Zone*. Vol. 1. Reprint, Germany: Verlag von J. Cramer, 1968.
——. *Trees and Shrubs of Mexico*. Washington, DC: Smithsonian Institution Publication 4461, 1961.
Staples, George W., and Derral R. Herbst. *A Tropical Garden Flora*. Honolulu: Bishop Museum Press, 2005.
Thulin, Mats. *Flora of Somalia*. London: Royal Botanic Gardens, Kew, 1995.
Tropicos.org. Missouri Botanical Garden. http://www.tropicos.org.
United States Department of Agriculture. Germplasm Resources Information Network. http://www.ars-grin.gov.
Van Wyck, Ben Eric, and Gordon Smith. *Guide to the Aloes of South Africa*. Pretoria: Briza Publications, 1996.
Von Maydell, H.-J. *Arbes et Arbustes du Sahel*. Germany: Eschorn, 1983.
Wagner, Warren L., Derral R. Herbst, and S. H. Sohmer. *Manual of the Flowering Plants of Hawai'i*. 2 vols. Honolulu: Bishop Museum Press, 1990.
Walter, Eric. *Echeveria*. San Francisco: California Academy of Sciences, 1972.
Weakly, Leigh. *Gardening in the Dry Tropics*. Townsville, Australia: Society for Growing Australian Plants, 1987.
Westly, Sidney B., and Mark H. Powell, eds. *Erythrina in the New and Old Worlds*. Nitrogen Fixing Tree Research Reports Special Issue. Pā'ia, HI: Nitrogen Fixing Tree Association, 1993.
White, F., F. Dowsett-Lemaire, and J. D. Chapman. *Evergreen Forest Flora of Malawi*. London: Royal Botanic Gardens, Kew, 2001.
Whitelock, Loran M. *The Cycads*. Portland, OR: Timber Press, 2002.
Williamson, Jessie. *The Useful Plants of Nyasaland*. Government Printer, 1955.
Winrock International Institute for Agricultural Development. *Nitrogen Fixing Tree List*. Little Rock, AR: Nitrogen Fixing Tree Association, 2005.
Worthington, T. B. *Ceylon Trees*. Colombo: Colombo Apothecarie, 1959.

INDEX

MAIN ENTRY PAGES ARE IN **BOLD**.

B

H

I

J

U

V

W

Y

Z

ABOUT THE AUTHORS

FRED D. RAUCH is emeritus professor of ornamental horticulture at the University of Hawai'i, where he served as state extension specialist in horticulture for twenty-five years. He served as adviser to the Hawai'i Association of Nurserymen and helped found and served as adviser to the Landscape Industry Council of Hawai'i. He served on the boards of the Friends of Foster Garden, the Halawa Xeriscape Garden, and was board member and president of the Western Region of the International Plant Propagators Society. He was honored with the 1994 Malama 'Aina Award by the Hawai'i Chapter of the American Society of Landscape Architects. Dr. Rauch grew up on a farm in a small town in Oregon and is a graduate of Oregon State and Iowa State Universities. He has been involved with the study of temperate and tropical ornamental plant material for almost sixty years and their photographs for about forty-five years. He currently resides in Oregon.

PAUL R. WEISSICH, A.S.L.A., is a licensed landscape architect whose familiarity with tropical landscape species has resulted in numerous consultant assignments. From 1957 to 1989, he was director of the Honolulu Botanical Gardens, where he expanded the two-garden system from 50 to 650 acres, covering four sites of differing ecological situations. He also increased the plant collection to a position of international recognition. Weissich coauthored *Plants for Tropical Landscapes* and *Small Trees for the Tropical Landscape* with Fred Rauch and *Na Lei Makamae: The Treasured Lei* with Marie McDonald. He also wrote *Majesty II: The Exceptional Trees of Hawai'i.*

PRODUCTION NOTES FOR

RAUCH AND WEISSICH /
PLANTS FOR THE TROPICAL XERISCAPE

Jacket design by Julie Matsuo-Chun

Text design and composition by Julie Matsuo-Chun with

display type in Gotham and text type in Georgia

Printing and binding by Regent Publishing Services, Ltd.

Printed on 128gsm matt art